HIDDEN WORLDS

HIDDEN WORLDS

HUNTING FOR QUARKS IN ORDINARY MATTER

TIMOTHY PAUL SMITH

PRINCETON UNIVERSITY PRESS
PRINCETON AND OXFORD

Published by Princeton University Press, 41 William Street,
Princeton, New Jersey 08540
In the United Kingdom: Princeton University Press,
3 Market Place, Woodstock, Oxfordshire OX20 1SY

Library of Congress Cataloging-in-Publication Data

Smith, Timothy Paul, 1960–
Hidden worlds : hunting for quarks
in ordinary matter / Timothy Paul Smith.
p. cm.
Includes index.
ISBN 0-691-05773-7 (acid-free paper)
1. Quarks. I. Title.
QC793.5.Q252 S65 2003
539.7′2167—dc21 2002074878

British Library of Congress Data is available

This book has been composed in Galliard and
Futura Condensed Regular

Printed on acid-free paper.∞

www.pupress.princeton.edu

Printed in the United States of America

1 3 5 7 9 10 8 6 4 2

- *Kristina* -

And Her Infinite Patience

with My Many Projects

- CONTENTS -

- FIGURES -

- ACKNOWLEDGMENTS -

I would like to express my thanks to the people at the University of New Hampshire, where I started this book, and at the Thomas Jefferson National Accelerator Facility, where I was doing my research at the time. Among them are Bill Hersman, John Calarco, John Dawson, Maurik Holtrop, Nathan Isgar, and many more who got me started on nucleon structure. I would also like to thank the people at the Massachusetts Institute of Technology–Bates Linear Accelerator Center, where I finished this book. These include faculty and co-workers Townsend Zwart, Richard Milner, Ed Booth, June Matthews, Karen Dow, Doug Hasell, Kevin McIlhany, and many more.

But it took more than just learning about physics and experiments to create this book. It took the ten thousand questions from students and fellow physicists who made me hone my description, and even my understanding, of the physics of nucleons and quarks. The list is long and includes Ben Yoder, Mark Szigety, Jeff Vieregg, Lisa Goggin, Vitaliy Ziskin, and Abby Goodhue, who work directly with me. Also Adrian, Peter, Adam, Aaron, Chris, Ben, Tong, Ehsan, Dan, William, Tavi, Nikolas, Hauke, Tancredi, and the rest of the army of curious minds who are working on these experiments.

I must also mention my family, who supported me through this project. First my boys, Will and Robin, who watched this book grow, not as fast as they grew, often in the early hours of Saturdays and vacations. Finally, Kristina, who as a physicist read, reread, and reread this book again, and as my wife encouraged and supported me with infinite patience. My deepest thanks.

HIDDEN WORLDS

- 1 -

Hidden Worlds: The Search for Quarks in Ordinary Matter

BY MOST accounts, the quest to understand the basic structure of matter has been an old-fashioned success story of growth and expansion. Machines have become bigger; computers have gotten faster. Beams of light or particles have become brighter and more powerful. Interactions of elementary particles have become more fleeting, and have given rise to ever more energetic and more exotic by-products. The basic engine driving this growth, the particle accelerator, began as a tabletop instrument you could hold in your hand—and was no more powerful than a lightbulb. By the 1950s, accelerators had grown large enough to fill a small warehouse, and drew enough power to run a large printing press. Now they need farmland or rangeland to accommodate their dimensions, and enough power to run a medium-size city.

Higher energies enable experimenters to "see" finer and finer details, to probe and analyze matter at smaller and smaller scales. Today the world's most powerful particle accelerators are operated at FermiLab, on the Illinois prairie about an hour's drive southwest of Chicago, and at CERN (Council Européen pour la Recherche Nucléaire—The European Organization for Nuclear Research), the particle physics laboratory, in a rural suburb of Geneva just under the Swiss border with France. The FermiLab accelerator is a ring 4 miles around; inside the ring there is plenty of room to pasture a herd of buffalo. At CERN, an even larger accelerator is under construction. When it comes into service in 2006, it will be 16 miles (27 kilometers) around.

At both these laboratories the combined accelerators and detectors are, in effect, magnificent microscopes that owe their magnifying powers to their ability to focus energy into electrons or protons that carry a trillion volts. With these energies both machines can resolve details in the structure of matter smaller than 10^{-18} meter across—a billionth of a billionth of a meter. And both machines are examples of science so big and so costly that they stretch the resources of individual sovereign states. CERN is funded by a European-wide consortium, and both laboratories are used by a collaboration of scientists from all over the world.

But the high price of the ability to probe such details is not measured only in dollars and cents. High energies can magnify, but they also carry great destructive power. It was often said in the early decades of high-energy physics that its basic investigative tactic was much like smashing two fine Swiss watches together in midair and then trying to understand how they worked by looking at the fragments. In fact, the true situation is even worse than that. Particles accelerated by today's state-of-the-art machines collide so violently that the collision fragments are often strikingly different from ordinary matter.

For many—even most—elementary particle physicists working today, that's just the point. The emphasis in the past few decades at such places as FermiLab and CERN has been to produce some of the most exotic particles predicted by theory, the "top" and the "bottom" mesons. On a different front, at Brookhaven National Laboratory on Long Island, a campaign is under way to create a "quark-gluon plasma," a state of matter, as some physicists have described it, "not seen since the big bang." Such phenomena can be studied only by accelerating, smashing, and, in effect, heating and squeezing ordinary matter to conditions beyond the edge of extreme: far beyond the temperatures and pressures prevailing even in the cores of the hottest stars, to regimes in which matter takes on strange and outlandish forms that do not exist at all in the universe as we now know it.

Yet in that rush to re-create such exotic conditions, and to study their implications for the birth, death, and ultimate structure of the universe, elementary particle physicists have almost forgotten the world in which we live. If the initial intellectual impulse was to probe the proton and the neutron in order to understand their role in ordinary matter, that impulse has virtually disappeared from the CERNs, FermiLabs, and Brookhavens of the world.

To my mind, that's a shame. I don't live at the dawn of time, and I don't live in a fantastically hot and energetic collision. I live in a world made up primarily of electrons, neutrons, and protons. And I want to know how they act and interact under "ordinary" conditions. I am a nuclear physicist, or to be more precise, a nucleon physicist. "Nucleon" is the generic term for neutron or proton, the particles that make up the atomic nucleus. My work and the work of my closest colleagues is dedicated to understanding the physics of ordinary nucleons, a layer in the onionlike organization of matter that gives rise to an incredibly rich set of phenomena. Those phenomena are quite literally destroyed among the debris of the highest-energy accelerators.

Ordinary Matter

Think about it this way: the universe that we understand is more than 99.95 percent neutrons and protons by mass. It is true that there are things we physicists don't understand, such as the stuff astronomers and cosmologists call dark matter. But the range of things we do understand in terms of electrons, neutrons, and protons is astonishing. Stars, those hot, glowing beacons in space, those light- and life-giving orbs suspended in the void, are made up of these three materials. Nebulae, the misty veils of interstellar gas and dust that stretch like curtains across the galaxy, are made of the same stuff. Even the most exotic of stars, the so-called neutron stars, are made of the same three ingredients.

Closer to home, the wind, the rain, and the earth beneath our feet are made of these three building blocks. Even the substance of life itself—the blood coursing through our veins, the brain and nerve tissue that provide the scaffolding for our thoughts, the deoxyribonucleic acid (DNA) that carries the blueprints from which each of us is built—is made up entirely of electrons, neutrons, and protons. Books and tables, hands, hearts, and heads, are made of these three most basic substances.

At one time, not so many years ago, the study of such particles coincided with the frontiers of high-energy physics. One of the first laboratories I ever worked with was the National Instituut voor Kernfysica en Hoge Energie Fysica (NIKHEF) in Amsterdam, The Netherlands. Even if you don't speak Dutch, you can probably understand most of the name: read "hoge energie fysica" phonetically and you will hear "high-energy physics." Not so obvious is "kern-fysica," which corresponds to the English phrase "nuclear physics." Still, whether one uses the term "nuclear" or "kernel," the word is meant to emphasize the role of the nucleus at the very heart of the atom.

That's an important clue to understanding what stirs the soul of the nuclear and the high-energy physicist. When the word "nucleus" was coined in 1912, it was viewed as the "atom" had been before it: as the ultimate, indivisible, fundamental particle of matter. The nucleus stood at the core or kernel of the atom, the sun about which the planetary electrons orbited. Nuclear physics, therefore, was essentially a quest to discover and describe the most basic building blocks of the universe. When physicists discovered that the nucleus itself was divisible into nucleons and had structure, and that those nucleons also had internal structure, the "dream of a final theory" of ultimate particles had to be abandoned within the domain of nuclear physics. That dream was passed on to high-energy physics, a discipline that, by its very name, no longer defined itself by assumptions about where the ultimate particles would be found.

The NIKHEF accelerator could accelerate electrons to energies of 770 million electron volts (MeV). A 770 MeV electron beam can probe matter at a scale slightly smaller than a fermi, or 10^{-15} meter. That is just powerful enough to resolve structure in the atomic nucleus itself; you could say that NIKHEF marks the start of nucleon physics.

Another laboratory that I visited recently is called DESY (pronounced "daisy"), the Deutsches Elektronen-Synchrotron, in Hamburg, Germany. DESY's beam energy is 30 billion electron volts (Giga electron volts [GeV]), 40 times more energetic than NIKHEF's, which also makes its resolution 40 times finer. The trade-off is that DESY's "field of view" is too small to be of much use for the study of nucleons. It could reasonably be argued that DESY marks the energetic upper limit of nucleon physics. The beam energy is so high that what you see are the "bare" constituents of nucleons. These constituents interact so violently that the nucleons themselves no longer maintain their identities, but become transformed instead into new and exotic particles. Indeed, most physicists at DESY identify themselves as high-energy physicists, not nuclear or nucleon physicists at all.

Beyond DESY, the high-energy frontier has moved even farther on, to CERN, to FermiLab, to Brookhaven, and elsewhere. NIKHEF and other accelerators of its class have long since been displaced as record holders for high energy, just as they had displaced their predecessors. But the continual historical advance of the high-energy frontier hardly means that the "captured" territories have been subdued, much less fully mapped or colonized. The physicists rushing ahead with plans for ever more powerful accelerators, for probing ever more deeply into the ultimate building blocks of matter, have seldom stopped to fully plumb the structures that they found along the way. Yet nucleons represent a level in the organization of matter having exceptional stability, unique in the universe. Many physicists, and I am one of them, want to know as much as possible about electrons, neutrons, and protons.

What Makes a Nucleon?

Just what do we physicists already know about these three particles? In the case of the electron, the summary can be quite brief. Electrons stand stark and innocent before us. What we see is essentially what we get: infinitesimally small particles, each with one unit of negative electric charge (−1), a spin of $\frac{1}{2}$, and a mass of 9×10^{-31} kilogram—some 2,000 times lighter than either the proton or the neutron.

And that is all. We can say no more about the physical properties of the electron because it appears that there is nothing more to say. I don't mean to be dismissive about the electron or its physics. Electrons are undoubtedly useful to twenty-first-century humankind, when we push or pull them through wires in the form of electricity. They are fundamental, of course, to the architecture of matter: the number of electrons and their orbital patterns in an atom or a molecule are what give rise to all of chemistry. Finally, the self-interaction of the electron is at the core of one of the greatest theoretical successes of the twentieth century: quantum electrodynamics, or QED. QED serves as a prototype on which theories of other particles (the quarks) are modeled. And QED has generated some of the most precise, experimentally confirmed predictions in all of physics.

But the electron itself appears to be truly elementary. Electrons seem to have persisted unchanged since the dawn of time, and they are likely to remain as they are, immortal, until the final sunset of the universe. There is no evidence that anything exists inside the electron; there are no "ultimate electrons" rattling around inside its shell. In all experiments ever performed it really and truly appears to be pointlike.

Not so the nucleons: the proton and the neutron. A hundred thousand times smaller than the smallest atom, both the proton and the neutron are measured in fermis. Yet as small as that might be, there is a world hidden inside each one. At first glance that world appears to be exceedingly simple. Protons and neutrons are each made up of exactly three particles known as quarks. Whatever quarks really are, just think of them for the moment as three balls rattling around inside of the sphere we call a proton or a neutron. There are two common kinds of quark, known as the up quark and the down quark. The proton is made up of two up quarks and one down quark. The neutron reverses those numbers: it is made up of one up quark and two down quarks. The fact that the proton and the neutron each have three quarks gives rise to striking similarities in their masses, sizes, and interactions. But that single difference, an extra up quark in the proton, an extra down quark in the neutron, also accounts for their unique characteristics: their differing electric charge, their decay patterns, and the details of how they couple.

The genesis of such profound differences merely out of varying combinations of simple parts should not be too surprising. Science has vast experience at larger scales with objects whose distinctive properties arise out of the number, identity, and arrangement of their parts. The familiar shorthand for specifying molecules takes tacit advantage of the fact that it is often enough just to enumerate their elemental components: H_2O—water—is two parts hydrogen and one part oxygen. Of course there are cases in which one set of atomic ingredients (such as $C_6H_{12}O_6$) can give rise to two or more distinct but related molecules (in this case, glucose and fructose). Chemists call them isomers.

When we ratchet up the magnification and view the world on the atomic level, the very identity of an atom—the definition of its elemental type—depends solely on the number of protons in its nucleus (which is equal to the number of electrons orbiting that nucleus, in the electrically neutral atom). Helium is helium because it has two protons and two electrons. Carbon is carbon because it has six protons and six electrons. Quantum mechanics dictates how those six electrons in the carbon atom arrange themselves physically, and that arrangement in turn ordains the chemistry of carbon: what it will bind to, with what strength, in what configurations. The concept of isomer has an analogue in atomic physics as well: a given element can come in several forms known as isotopes, chemically almost indistinguishable from one another, but different nonetheless in the number of neutrons that share real estate in the nucleus with the protons.

In sum, it seems entirely natural that the properties of the proton and the neutron themselves arise from the number and arrangement of the quarks that make them up. One of the chief burdens of this book is to show how their properties and configurations give rise to a "quark chemistry" of remarkable complexity—just as the configurations of atoms in molecules and the configurations of electrons in atoms give rise to ordinary chemistry. The concepts of isomer and isotope, for instance, have an analogue in the world of quarks, as we shall see with such particles as "Δs" and "Ropers."

In Vitro versus In Vivo

But before exploring those details, the very idea of quark chemistry can shed light, I think, on a curious question that afflicts nucleon physics with far more confusion than seems necessary. The question goes to the heart of the discipline: just what is nucleon physics? I once posed this question to a number of my colleagues during lunch at a conference at a laboratory that specializes in what I call nucleon physics, the Jefferson Lab, in Newport News, Virginia. And I was surprised to find that even my use of the phrase "nucleon physics" was controversial. Some people wanted to call it "intermediate-energy physics," a name motivated partly by the kinds of accelerators the work relies on and partly by the recognition that the study of the quarks inside nucleons bridges the study of the role of nucleons in the atom (nuclear physics) and the study of high-energy physics proper.

Yet most of the physicists at our lunch saw what we nucleon physicists do as less of a branching away from nuclear physics than as an extension of it—and an obvious extension at that. Further muddying our discussion is the way physicists identify themselves: who is a nucleon physicist and who is a high-energy physicist? Two experimentalists working side by side on the same experiment might identify themselves as "nuclear" or "high-energy" physicists, not on the basis of the experiment at hand, but rather on the basis of their own experiences in graduate school. Physicists who had been part of a high-energy research group in graduate school might wear that tag for an entire career. Likewise, physicists whose advisors called them "nuclear" could carry that label into retirement.

Perhaps it is not surprising that the labeling of physicists is a social, hence largely accidental, phenomenon. But the labeling of the subdisciplines of physics need not be the product of historical accident as well. Part of the confusion about nucleon physics stems from the historical shift I noted earlier in the boundaries of high-energy physics. A 1 GeV electron accelerator in the 1960s clearly belonged to the high-energy community, whereas at present it is the province of the nuclear physicist. What that early convergence in labeling concealed was that the two fields have named themselves according to different criteria. Nuclear and nucleon physics are named after the subjects they study; high-energy physics is named after the machines and detectors it uses. If nucleon physics was once "high-energy" physics, but the high-energy frontier has now moved on, doesn't it follow that nucleon physics has now become "intermediate-energy" physics?

The answer is no. But let me try to clarify the situation by moving from a chemical analogy to a biological one. In biology, laboratory workers often remove and purify a cell line to study the cells "in vitro," that is, in a test tube. ("In vitro" literally means "in glass.") In vitro studies can give "clean," unambiguous results. Any variables—temperature, radiation, chemical concentrations—that might affect the results can be scrupulously controlled. But no biologist would jump to the conclusion that some result observed in a test tube or petri dish implies that the same results would necessarily unfold "in vivo." ("In vivo" means "in life," that is, in the living organism.) The effect of a drug on cells in vitro, for instance, might be quite different from its effect on those same cells in their natural habitat, the body of the organism from which they come. The behavior of cells in vivo is often so complex and so different from their behavior in vitro that in vivo study is a complementary discipline in its own right.

Once I was at a computational workshop at FermiLab when I was asked what distinguished the studies of protons, neutrons, and their quark constituents that we pursue from the studies that high-energy physicists make of the same particles. My answer was "in vivo versus in vitro." The proton and neutron physics we study is the study of quarks in vivo, in their natural home within the body of a nucleon. High-energy physics looks at quarks in a more rarefied, extra-nucleon environment. It turns out that nature forbids the observation of a single quark, but you can infer a great deal about the nature of quarks from the fragments they form. To pursue the Swiss watch analogy, nucleon physics studies the watch by listening to its ticks, and, without opening it, tries to infer what is going on inside.

So I propose the following working definition of the differences between nucleon and high-energy physics:

> High-energy physics is the study of quarks, gluons, bosons, and other exotic, fundamental, and elementary particles of nature, whereas nuclear/nucleon physics is the study of these particles as they combine to form normal matter; as they combine within the hidden world buried inside protons and neutrons.

Confinement

No matter how you study quarks, one of their most distinctive properties makes investigating them a peculiar and difficult exercise: quarks are always hidden, buried deep within some larger particle. No one has ever been able to isolate a single quark. It is not just that we have not been clever enough to build a "proton smasher," or some better machine or experiment. Rather, nature has contrived its laws in such a way that not only have we never seen an isolated quark—but we never will! That is the root of a great deal of frustration, and at the same time the source of the scientific challenge that we explore in this book.

So how do we proceed? Shouldn't this "non-observability" lead to a kind of intellectual crisis? Doesn't the forward movement of science have to be fueled by observation? Yet—it bears repeating—*we will never observe quarks.* I suspect that a staunch positivist would have a field day with the quark hypothesis. We cannot see them, or even detect them, yet we know a great deal about them. To overcome that limitation we will have to examine and rethink what is meant by good evidence. In fact, a special emphasis of this book is to consider how we know what we claim to know.

A lot of books about contemporary science report the end products of a generation of research. A book on quarks would give great prominence to the hard-won list of their six known "flavors." But for me the road to these results is at least as interesting as the results themselves. The issue is one of personal taste and appreciation, but let me try to explain it this way: I admire the great beauty of a medieval cathedral such the Dom, in Graz, Austria, or York Minster, in northern England. But when I think about how these vaulting expressions of the human spirit were handcrafted in a preindustrial age, without girders of iron or steel, the cathedrals are transformed to my sensibilities from the merely beautiful to the truly magnificent.

So how can we know anything about something we cannot even see or sense? Briefly, the solution is to make theory and experiment work so closely together that they become interleaved. Given a theory of quarks, how might a proton be built out of them, and what, under various testable conditions, would we see? Then, when we do see something that roughly matches those expectations in our experiments, we promote and fine-tune the theories that predicted it. These general procedures hold for any attempt to reconcile theory with experiment. Yet our experiments that look at the way quarks combine to build particles are not like the ones conducted at the high-energy laboratories such as the Stanford Linear Accelerator Center (SLAC) in California, or FermiLab, or CERN. Instead of hurling particles at each other with such fury and energy that they are destroyed, we tickle and excite the particles. In that way we study the quarks "in vivo."

The Secret Lives of Quarks

The laboratories that measure the shape and excitations of protons and neutrons "in vivo" are of modest proportions, compared with high-energy facilities. The energy of the electron beam at Jefferson Lab can reach roughly six billion electron volts. MIT-Bates Lab, north of Boston, has an accelerator capable of delivering particles whose energy reaches a billion electron volts, 1,000 times smaller than the "Tevatron" at FermiLab. But these laboratories do have unique features. They deliver high currents with high precision, and they can manipulate the alignment and orientation of the spins of the electrons, protons, and neutrons. By using these so-called polarization tools they can disentangle the nuclear effects from the quark effects, nudging out the secrets of the proton and the structure of the neutron.

"A riddle wrapped in a mystery inside of an enigma." Winston Churchill's words could apply to quarks as well they did to the old Soviet Union. What are these most lilliputian particles that lie forever hidden from our gaze? What are we trying to measure in our nucleon physics laboratories? One curious property of protons and neutrons is that they vibrate at certain fixed, resonant frequencies; they have a number of natural "harmonics." We will talk about how to "ring" a proton—that is, hit it hard enough to make it resonate. That very concept brings us back to the interplay between theory and experiment, which is so important for unraveling the riddles of the quark. A good theory should explain how hard we need to strike a proton to get it to ring. A good theory should be able to predict the magnitude, shape, and energy of a resonance. It should also explain why such resonances exist, and then go a bit further,and predict what no one has ever seen. At the same time, a good experiment should be able to distinguish among an entire spectrum of candidate theories, identifying the good ones and rejecting the ones that diverge from the data. On occasion an experiment should even show us something the theorists had never thought of modeling with their theories. Sometimes an observation is completely unexpected.

Resonances are just one test, only one of the kinds of clues we have to solve our mystery. We can also ask, What is the shape of a neutron or a proton? Are the quarks inside them rigidly fixed in space, or do they float about freely? Or perhaps the truth is something in between: do the up quarks tend to congregate in one region of the nucleon and the down quarks in another? How do these quarks orbit and swirl around each other?

Three quarks, sometimes exhibiting a resonance, sometimes performing a quantum mechanical dance—is that all there is inside a proton or a neutron? Yes and no. When we peek just beneath the surface of the proton, all we see is this simple choreography of three quarks. But on closer inspection something else is going on. There seem to be emissaries dashing back and forth between the stately quarks. The intermediate particles are called gluons. They make the quarks aware of one another, and, as their name implies, they bind the quarks together. These gluonic emissaries can add their own jigs to the dance, giving rise—perhaps—to a new resonance, a new harmonic "ring."

After watching the ballroom floor for a while, we start to notice other dancers. Wasn't there another couple of quarks swirling around each other off in a corner, just for the briefest moment? They seemed to pop into existence and then vanished again. Yet somehow they did it without defying that cardinal rule: three quarks and three quarks only on the dance floor.

To understand how we can see this quantum choreography, this stately waltz, we need to discuss three major tools of the trade. How can accelerators push electrons nearly to the speed of light? (Only then do they have enough energy to ring or tickle a nucleon.) Curiously enough, these fascinating machines are built essentially out of microwaves and magnets. How do we detect particles such as protons and neutrons, which are a quadrillionth of a meter across, or electrons, which are even smaller, with little more than charged wires and plastic that glows? And how can we find our way through a terabyte of data (a terabyte is a million megabytes), and then say that we "saw" the shape of a neutron, or "heard" the delta resonance, or "felt" the vibrations of the dancing gluons?

The evidence is indirect, and so we need to proceed deliberately, one step at a time. But the idea that within my lifetime a whole new world has been glimpsed inside of the proton and the neutron is intriguing and exciting, an intellectual adventure of the highest degree. "Three quarks for Muster Mark!" wrote James Joyce in *Finnegan's Wake*, with seeming prescience. By the time he wrote that last novel, Joyce was nearly blind. Now we too will find out how much we can get to know without really being able to see.

- 2 -

The Rise and Fall (for the right reasons) and Rise Again of the Quark Hypothesis

IN PHYSICS TEXTBOOKS we are told that quarks were proposed in 1964 by Murray Gell-Mann of Caltech and George Zweig of CERN, and experimentally verified by a team of physicists led by Henry Kendall and Jerome Friedman of MIT and Richard Taylor of Stanford in 1972. It is an account that mimics the way physicists think their field should progress: a theoretical prediction followed by experimental verification. What is glossed over in this tale is that the quark hypothesis should have been rejected in the late 1960s based on some very good experimental evidence.

No matter how compelling a hypothesis or theory is, in the world of science it really must stand up to the scrutiny of experimental evidence. In the case of quarks, the whole notion of what is good evidence is very tricky. No one has ever seen an isolated quark, and there is good reason to think that no one ever will. Still, after more then three decades of experimental and theoretical investigation, we know a great deal about them. We know that they have "spin" and come in different quantum "colors" and "flavors," and possess other characteristics that will be discussed in this book.

The fact that we know anything about something so small is in itself amazing. A neutron or a proton is 10^{-15} meter across, and quarks are even smaller. To give an idea of what that scale means, there is roughly a factor of 100,000 between the size of a person and the size of a cell. There is another factor of 100,000 between the size of a cell and the size of an atom, and one final factor of 100,000 between the size of an atom and the size of a neutron. And all experimental evidence tells us that quarks are significantly smaller.

However, one might argue that the discussion of quarks by themselves is a bit academic. Quarks are always in bound states—in confinement—and it is this confinement that lies at the heart of the story of the near rejection of the quark hypothesis. In one sense confinement is just the statement of the observational fact that we have never seen a free quark, that they are always confined inside protons or neutrons, or some other particle. On the other hand, confinement is perhaps the most unique feature of the quark world. The neutron and the proton cannot be described simply as scaled-down atoms or tiny solar systems. Rather, they are something whose elements we can never really hope to pull apart. That is not to say we cannot unravel some of their mysteries. It is just that the techniques will be a good deal different from simple dissection. Still, confined within the proton and the neutron, and many more exotic particles, is a world with its own etiquette and landscapes. We know something about the rules by which quarks order and group themselves, and we are just learning to decipher and chart the choreography of quarks as they orbit and dance around each other.

It was a long road from the days of splitting the nucleus into protons and neutrons to the point where physicists were ready to accept the fact that the constituents of the next smaller scale were, in the old sense, beyond their reach. It took time to accept the concept of confinement, but this abstract concept was always driven by good, solid experimental evidence. In reality, the evidence that eventually persuaded the particle physics community to accept the quark hypothesis did not arise from a single line of investigation. Rather, it was the result of two different lines of study, two separate threads. The first thread followed the discovery of a multitude of elementary particles in the 1950s. The second thread followed the probing of the structure of the proton.

In the late 1950s, the field of subatomic physics had become a crowded place, with new particles being discovered almost weekly. The world was not populated only by the familiar electron (e^-), proton (p), and neutron (n), but now also muons (μ^-), pions (π), τ, neutrinos (ν), ω, Λ, ρ, η, and Ξ. The Greek alphabet was nearly exhausted! One of the chief features of many of these new particles was a new property called "strangeness." In the 1950s, this property was exactly what its name implied—strange. Experimenters looked at photographic plates, which had been exposed to cosmic rays on mountaintops or during high-elevation balloon or airplane flights. The photographic plates were sensitive to charged particles passing through them, and on them physicists would see the apparently spontaneous production of a proton and a pion (π^-) (see figure 2.1). They inferred that some type of neutral particle, which left no track, must have decayed. It could not be a neutron (the neutron did not have enough mass, or energy, to create a proton and a pion)—so they named the particle the Λ, because of the inverted *V* that they saw on their photographic plates.

This same particle was soon observed at accelerators that slammed high-energy protons into targets and produced a shower of particles. But it was not the only "strange" particle. The Ξ, the kaon (K), and the Σ were also observed, each with its own unique mass and other characteristics, and eventually each of them decayed in a pattern similar to the "strange decay" first observed with the Λ.

When the world was just made of the electron, proton, and neutron, life seemed simple and complete. But with the arrival of so many newcomers the question became a problem of taxonomy: how do we organize all of them? Experimenters had been able to measure lifetime, masses, charge, and decay patterns, and now they dearly desired to create some type of classification scheme, reminiscent of Dmitri Mendeleev's *Periodic Table of the Elements*, the most basic table for chemistry.

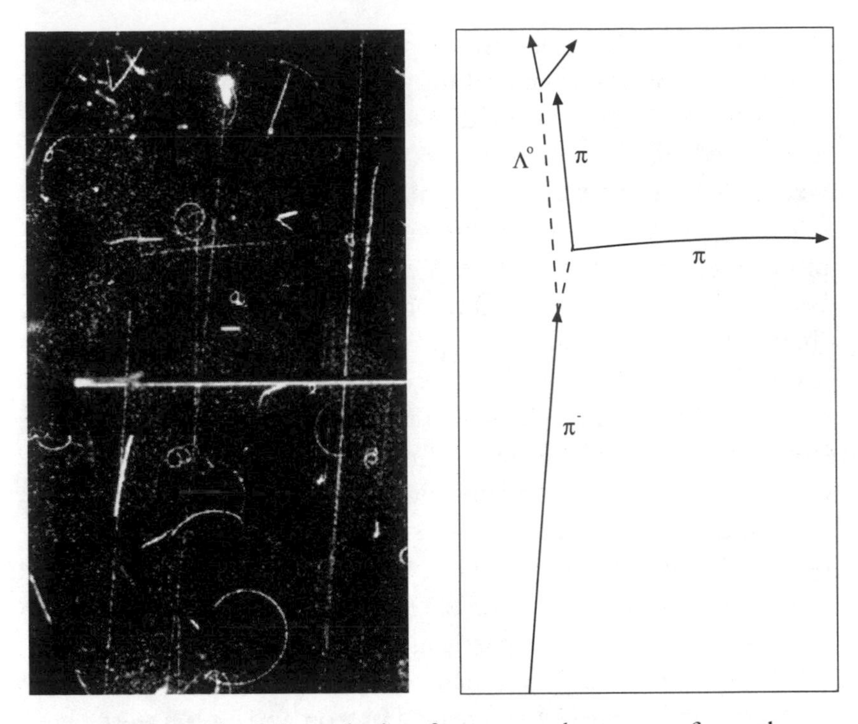

Figure 2.1 Much of the early data for strange decay came from photographic plates like this one. Charged particles leave tracks in the plate, and the curvature indicates the charge and momentum of the particle. (Courtesy of Brookhaven National Laboratory)

The periodic table has embedded in it the structure of the atom and even the two spin states of an electron. This is remarkable since Mendeleev first compiled the table in 1871, decades before the discovery of the atom and half a century before the quantum description of its structure. He ordered the elements first by increasing weight and also started a new row when physical characteristics of elements repeated. Thus, hydrogen, lithium, sodium, potassium, and so on are all in the first column or "group" because chemically they are very similar. The last column contains the noble or inert gases: helium, neon, argon, krypton, and so on. The structure of the atom that creates these chemical characteristics is the electron orbit. The first row in the table contains two elements, because the first orbit can contain up to two electrons. The second row has eight elements, and the second orbit of the atom can have up to eight electrons. The elements in the last column, the noble gases, all have their orbits completely filled and so are chemically inert. The elements in the first column, the alkali metals, have only one electron, which is easily oxidized or ionized, in their outermost orbit. Thus by arranging the elements according to the right characteristics, the periodic table hinted at their underlying structures.

In 1961, Murray Gell-Mann organized the recently discovered particles in a paper entitled "The Eight Fold Way: A Theory of Strong Interaction Symmetry." Strongly interacting particles are those that bind in the nucleus, and include all the particles listed above except the electron and its exotic heavier siblings, the μ and the τ, and the neutrinos (ν). "The Eight Fold Way" sounds Taoist—as it was meant to sound—but the title truly reported the results of identifying the right symmetries or characteristics that are used to classify all these new particles, and would hint at the underlying quark structure.

Gell-Mann wanted to plot the particles on some type of graph. One might try to list the particles by mass, but many particles would sit almost, but not quite, on top of each other; for example, the proton and the neutron have masses of 939.5 MeV/c^2 and 938.2 MeV/c^2 (about 1.7×10^{-27} kg or 3.7×10^{-27} lbs), a distinction of only 0.1 percent. Perhaps they could be spread out by placing charge along another axis? (See figure 2.2.) But no useful patterns emerged. Perhaps that should not be surprising, since weight by itself had even upset Mendeleev's periodic table originally. Argon was more massive than potassium, but the characteristics of argon put it just before potassium. When the periodic table was readjusted by atomic number instead of by atomic weight, the patterns cleared. For Gell-Mann, the key would be turning to the decay patterns.

Decay patterns tell us about quantum number. The premise of this argument is that particles have two types of properties. They have transitory properties, like momentum or position, which are continuously changing, and more consistent properties, such

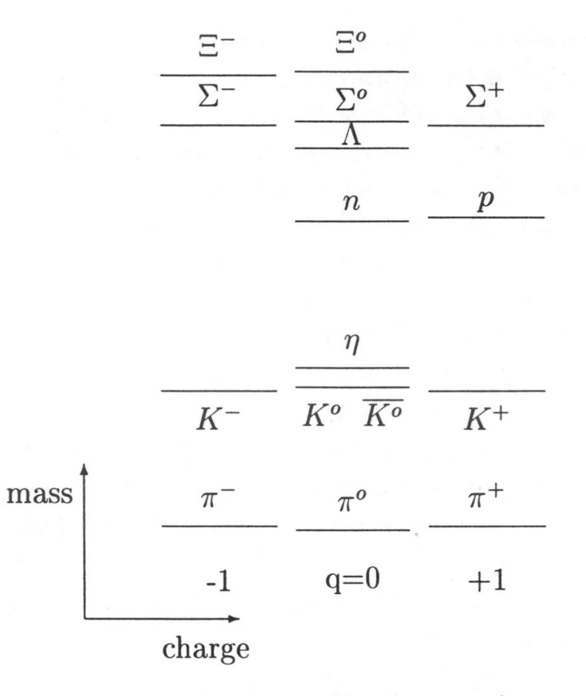

Figure 2.2 Particles arranged by charge and mass.

as charge (or even atomic mass), which resist changing. If a particle decays freely and quickly by a particular mechanism or process, then no quantum number has changed. However, if a decay takes a long time, and so a particle is relatively stable, then that decay changes a quantum number. For example, the particle N(1440), the "Roper" particle, will decay into a neutron in about 10^{-24} second, which is the amount of time it takes light to cross a proton! So we say a Roper is unstable, and in that decay no quantum number has changed. A proton has never been seen decaying into a positron ($p \rightarrow e^{+} \bar{\nu}$), although there is enough energy to allow this to happen. We say that a proton is stable—and quantum numbers such as "baryon number" and "charge" are conserved. Now in the case of a strange particle, such as a Σ, it can decay into a Λ and a π quickly (10^{-20} second or faster), but the decay of the Λ is rare—10^{-10} second, or 10 billion times slower. The first decay passes strangeness from the Σ to the Λ, but

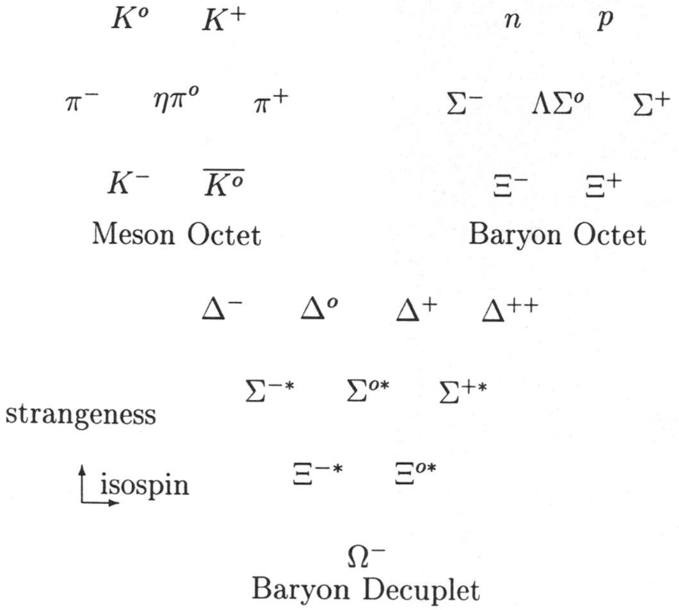

K^o K^+

π^- $\eta\pi^o$ π^+

K^- $\overline{K^o}$

Meson Octet

n p

Σ^- $\Lambda\Sigma^o$ Σ^+

Ξ^- Ξ^+

Baryon Octet

Δ^- Δ^o Δ^+ Δ^{++}

Σ^{-*} Σ^{o*} Σ^{+*}

strangeness

isospin

Ξ^{-*} Ξ^{o*}

Ω^-

Baryon Decuplet

Figure 2.3 Particles arranged by isospin and strangeness. The baryon octet, which includes the neutron and the proton, inspired Gell-Mann to call this the *eight-fold way.*

then the second decay is very slow. The Λ is the lightest strange particle, so when it decays, strangeness will be lost and the strangeness quantum number will go to zero. In later years we would understand this in terms of the decay of a strange quark—but that is getting ahead of ourselves.

Two quantum numbers that had been determined at the time for all particles were strangeness and isospin. Isospin plus is a measure of how "protonlike" the particle is, whereas isospin minus is a measure of now "neutronlike" it is. When Gell-Mann plotted all the particles on his isospin-versus-strangeness graph, several patterns emerged for various types of particles (see figure 2.3). The most important pattern, for most of the matter in our normal world, is the one that includes the proton and the neutron, the baryon octet or group of eight particles (hence the title of Gell-Mann's paper). There were two important consequences

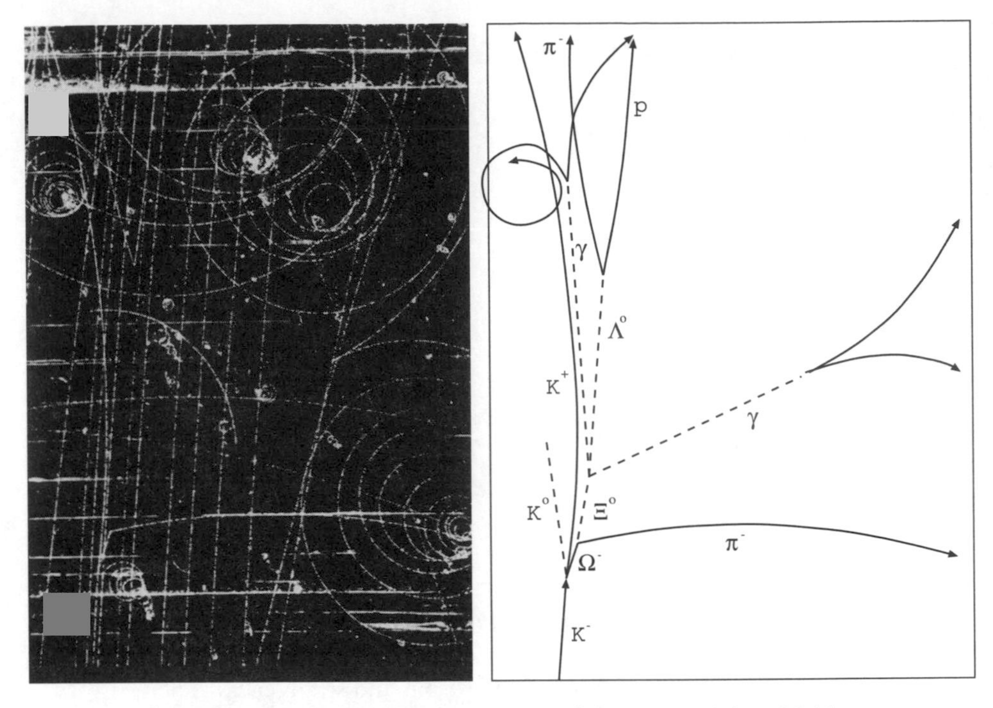

Figure 2.4 The first observation of the Ω^- particle—1964. (Courtesy of Brookhaven National Laboratory)

of these patterns. First, in the baryon decuplet, or ten-particle group, the Ω had not been observed. So these patterns predicted a new particle, and by extrapolating the masses of other particles in the decuplet, they were able to make a reasonable prediction of the mass of the Ω. In 1964, the Ω was observed in good agreement with predictions (see figure 2.4).

The second feature is that the patterns that emerged were recognized by mathematicians and theoretical physicists as belonging to something in group theory called "su(3)"—special unitary three. The 3 refers to the three bases or building blocks of the group. In mathematics the bases have a very clear role. There are certain and well-prescribed ways in which one can combine the bases to create new objects. Still, it was not clear whether these three bases or building blocks corresponded to anything physical, and if they did, what it was that they might be. People tried to develop schemes where the bases might be the proton, neutron, and Λ particle. There was even something referred to as the "boot-strap" theory, where the bases are a hybrid of all the particles, and all particles are created from these bases. It may seem cyclic—similar to lifting yourself up by your own bootstraps. Boot-strap theories were also referred to as "nuclear democracy," since all particles were equally important.

Finally, in the critical year of 1964, quarks were proposed twice. In January, from Caltech, Gell-Mann submitted a letter to a journal for publication, in which the term "quark" first debuted. That same month, in CERN, near Geneva, George Zweig wrote a paper on the possibility of subconstituents of protons and neutrons. Zweig was a "post-doc" at the time, which meant that he had just finished his Ph.D. and was viewed as being "green" and the most junior of scientists. His term for a quark was "ace," as in the phrase "ace in the hole," which has nearly been forgotten. Both papers, however, made two startling statements. Each paper included, first, the prediction of fractional charges for the quarks, and second, the prediction that one could observe quarks in normal matter:

> It is fun to speculate about the way quarks would behave if they were physical particles of finite mass (instead of purely mathematical entities as they would be in the limit of infinite mass). . . . Ordinary matter near the earth's surface would be contaminated by stable quarks as a result of high-energy cosmic ray events throughout the earth's history, but the contamination is estimated to be so small that it would never have been detected. A search for stable quarks of charge $-\frac{1}{3}$ or $+\frac{2}{3}$ and/or stable di-quarks of charge $-\frac{2}{3}$ or $+\frac{1}{3}$ or $+\frac{4}{3}$ at the highest energy accelerators would help to reassure us of the non-existence of real quarks. *Murray Gell-Mann, Phys. Letts. 8, p. 214 (Feb. 1, 1964)*

Gell-Mann and Zweig both felt that the eight-fold way and su(3) had driven them to postulate the quark and the $-\frac{1}{3}$ or $+\frac{2}{3}$ fractional charge. They also knew that there were no candidates for such a particle. So only with great reluctance did they publish their hypotheses. In later years Gell-Mann confessed that he chose the journal *Physics Letters*, because he felt that such a radical idea had a better chance of getting accepted and published in a journal that specialized in quick reviews and printing. Zweig published his article in a CERN in-house report, and then, as a junior theorist, was not able to publish it elsewhere for years.

One point on which Gell-Mann and Zweig differed was the question of the physical significance of a quark. From the beginning Zweig saw them as real particles—the physical building blocks of protons, neutrons, Λs, and so on. In his view, most decays could be explained by the rearrangement of quarks, and then the reluctance of the Λ to decay is because the strange quark really must decay and turn into something else—or at least into a different type of quark. Gell-Mann was more guarded, and he usually referred to quarks as "purely mathematical entities" for nearly a decade. He often referred to his hypothesis as a useful mathematical model that should be discarded once the desired results were obtained. In fact, his promotion of the search for fractional charges was in part prompted by a desire to discredit the physical quarks.

So what is a fractional charge and why is it radical? Ever since Millikan had published his paper in 1910, the charge of an electron or a proton seemed to be unique and singular. There was no evidence of any other charge, except for multiples of the charge on an electron (or proton), in the long list of known physical phenomena. In the opening paragraph of Millikan's famous paper, he writes:

> Among all physical constants there are two that will be universally admitted to be of predominant importance; the one is the velocity of light . . . and the other is the ultimate, or elementary, electrical charge. *R. A. Millikan, Phil. Mag. (series IV) vol. 19, p. 209 (Feb. 1910)*

That prejudice of "predominant importance" still stood half a century later.

What Millikan and his co-worker Mr. Begeman did in what every physics student now knows as the "oil drop" experiment was to create a mist of water droplets in a chamber. They then irradiated them with a radioactive radium source, thus knocking out electrons and leaving some of the atoms in the droplet ionized. Next they let the droplet fall while observing it with a microscope, and determined its terminal velocity, from which they could calculate the mass of the droplet. Then they turned an electrostatic field on, with the negative up and the positive down. Since the ionized droplet carried an excess of protons, there would be an electrostatic force upward, usually raising the particle up against gravity. By balancing the electrostatic force against the gravitational force, Millikan could deduce the charge on the droplet.

The experiment is extremely tedious. Trying to generate these clouds of droplets, or artificial fogs, by varying the pressure within the experimental chamber. Then trying to ionize a droplet before the whole cloud fell under gravity. Then watching a droplet fall, timing it, raising it, and balancing the forces on it, all while watching it through a microscope, and all before it evaporated. It is no wonder that in Millikan's paper he only reports measuring thirty-three drops total. He also wrote about one peculiar droplet:

> I have discarded one uncertain and unduplicated observation apparently upon a single charged drop, which gave a value of the charge on the drop some 30 per cent lower that the final value of e. *R. A. Millikan, Phil. Mag. (series IV) vol. 19, p. 209 (Feb. 1910)*

Some searchers for free quarks have cited this, in jest, as the first observation of fractional charge. More important to the serious search for fractional charged particles, the Millikan technique inspired a series of experiments from the mid 1960s through the 1970s. These updated experiments used for their droplets a variety of materials: water, mineral oil, graphite, iron, and niobium. By using the oil or solid droplets, these experiments solved one of Millikan's persistent problems: his water drops would evaporate and change their mass through the course of a measurement. Experimenters also used materials with tailored magnetic and dielectric properties, together with complex feedback loops that automatically balanced the fields. At one point, George LaRue, William Fairbank, and Arthur Hebard at Stanford had one niobium sphere that consistently yielded a $\frac{1}{3}$ charge measurement. This created a great deal of excitement at the time, packing auditoriums at conferences. But it stood alone as an isolated data point. After an extensive study of the systemic error of their experiment, they finally retracted their claim to the observation of a fractional charge. As sophisticated as the modern experiments were, with specially designed electrostatics and stabilizing magnetic fields, still no fractional charges were observed.

People were also looking for fractionally charged particles—quarks—in cosmic ray studies and at accelerator laboratories. Mounting an experiment is a very involved process, often taking years of preparation. Yet within a few months of the publishing of the original theoretical papers about quarks, physicists were "mining the data." For example, in an experiment designed to search for something like the Ω particle, or to measure the distributions of protons in carbon, or a hundred other things, there are hundreds of thousands of events (like the photographic plate shown above) that do not contain an Ω, but are still recorded. "Mining the data" means going back to the archives of data already taken and looking for the fractionally charged particle. It was a long and tedious task to pore over thousands and thousands of photographic plates, and still no evidence was found for a free quark in the archives of any laboratory.

The accelerator and cosmic ray experimentalists then mounted new experiments specifically designed to look for fractional charges. One of the traditional methods of measuring the charge of a particle is to see how its trajectory curves in a magnetic or electric field. The problem with this technique is that it really measures the ratio of charge to momentum or mass (depending on whether an electric or a magnetic field is used). If the particle is a known particle—such as a proton or electron—then we know its charge and even its mass, so we can calculate momentum, velocity, and even energy, given a map of its trajectory. But for an unknown particle type, knowledge of the charge to mass ratio alone tells us little. Either some assumptions are needed—such as the mass of the quark is one-third of the mass of a proton—or we need to make more measurements.

Two additional measurements that can be made on an unknown particle are the energy of the particle and the rate at which it loses its energy. The energy can be measured by letting the particle collide with a special material called "scintillator." Scintillator converts the energy of the particle into light. The amount of light is proportional to the energy of the particle, and is easily detected and measured. Scintillators actually are a whole range of materials that include both liquids and solids. An experimenter will choose the material depending upon its detection efficiency, response time, and, of course, cost, together with the expected energy of the particle. Typical plastic scintillator looks like Plexiglas until it is exposed to particles or radiation. In the presence of an ultraviolet light (such as a "dark light") the Plexiglas is unchanged, but the plastic scintillator will glow blue.

To measure the rate at which energy is lost one can construct the detector with several layers of scintillator. The total light from all layers is a measurement of the total energy of the particle, but by looking at how much light is in each layer, one measures how fast the particle loses energy—or how much it interacts with the material. The interaction with material is primarily an electric charge interaction. So a particle that loses its energy quickly has a greater charge than a particle with a small charge. Actually, even particles with no charge (such as neutrons and the ultraviolet light mentioned above) lose energy, but at a very different rate.

In fact, by just knowing the energy and the rate of energy lost, experimentalists could measure charge. Therefore, they built detectors with layers and layers of scintillator. Some of them pointed skyward to detect cosmic rays. Some of them pointed toward the targets of an accelerator laboratory. The advantage of cosmic rays is that they are everywhere, and they are free (of cost)—but the count rate is low. Accelerators have the advantage of very high rates, and they can be tuned to produce particles at a controlled high energy—but they are not free. However, despite considerable expense, a rush of activity, and a flood of new data, there were still no free quarks.

With hundreds of experiments performed worldwide throughout the late 1960s and early 1970s, and absolutely no evidence of a free quark, it would have been a reasonable and rational conclusion to reject the physical quark hypothesis completely. Yet it persisted for two reasons. First, it was still a solution to the problem that had originally prompted it: it explained all the particles seen in the 1950s, and it explained the symmetries described in the eight-fold way. Second, there was mounting, although unclear, evidence that there was something going on inside the proton.

The second thread or line of investigation that contributed to the evidence that persuaded physicists of the existence of quarks dates back to the early 1950s, when physicists at Stanford had been studying the proton with electron scattering. In 1955, a group led by Robert Hofstadter had scattered electrons off a hydrogen (proton) target, and measured the size of the proton to be roughly 10^{-15} meter, or 100,000 times smaller than the size of an atom. That is small, but it is not zero. The natural question that physicists were asking themselves was, What is the structure of the proton?

The experiments that searched for fractional charge or exotic particles were in some sense searches for anomalies. What was measured were the properties of individual particles. If any one particle was observed to have the sought-for properties, then the search was considered a success, be it fractional charge or strangeness = −3 (the Ω particle). Scattering experiments, in contrast, are statistical and indirect. In Hofstadter's, and subsequently all Stanford (later SLAC) experiments, they scattered an electron off a target and measured at what energy and in which direction the electron emerged. Note that here, the particles to be studied reside in the target, and the electron is merely a probe. The initial energy of the electron will determine the resolution of the probe—the smallest thing it can observe. A low-energy electron will have a long wavelength and low resolution, whereas a high-energy electron will have a short wavelength and a higher resolution.

If an experiment uses a beam of electrons with 1 MeV of energy, the electrons have a wavelength of 2×10^{-13} meter, or one five-hundredth of the diameter of an atom. Therefore, atoms are distinguishable, but the nucleus (which is 200 times smaller) is still seen as a point that carries the sum of the charges of all the protons in that nucleus. The electron will scatter exactly as if it was scattering from a pointlike object with charge $+Z$. At a higher energy—for example, 100 MeV—the wavelength is 2×10^{-15} meter, or 2 fermis. This is the scale of the nucleus, and we begin to resolve the structure of the nucleus. If our target is a heavy element with a lot of protons and neutrons, the electron will scatter not from a single point, but from a spatially distributed charge. (We will discuss electron scattering in more detail in later chapters, since it plays a critical role in the gathering of evidence about quarks.)

In 1955, Hofstadter's group was using Stanford's new accelerator with a 550 MeV electron beam on a hydrogen target. At these energies a hydrogen target is essentially a proton target. What they saw was that the charge of the proton was spatially extended over a region of space 0.8 fermi across. At the time many people were satisfied with the notion that the proton had a finite size, but there was always that question in the back of physicists' heads: "We looked inside the atom and saw the nucleus, we looked inside the nucleus and found the proton and the neutron. If we look inside the proton or the neutron, what will we see there?"

Prompted by the measurement of the finite size of the proton, and driven by the desire to see inside it, in 1957 Stanford University proposed the building of a newer and much larger accelerator. This accelerator and laboratory became known in the 1960s as SLAC, and delivered a 20 GeV beam to a target. The accelerator itself is over 2 miles long. At the time it was the largest and most expensive physics tool in the world, designed to look at the smallest scales—a resolution of about one one-hundredth of the diameter of a proton.

What experimentalists saw defied any expectations based on old models of the proton. The proton is not just a ball of charge—there is something, rather, there are many things, within it. However, it was still a long road to understanding what that structure is. Remember that the data can only be of the form "we saw n electrons scattered in this direction and energy, and m electrons scattered in that direction and at that energy." The experiments cannot see directly anymore without extensive theoretical interpretation. Separately from the experiments, theorists start from a model and calculate what the experiments would see if that model were true.

J. D. Bjørken was a young theorist at SLAC who turned his attention to these bizarre results. He believed that it was unlikely that protons had subconstituents, and he had at his fingertips the skills to test this hypothesis. Based upon an abstract branch of theoretical particle physics called "current algebra," Bjørken calculated certain features of the scattering that would be expected if protons and neutrons had subconstituents. One of the first predictions was the idea that the number of electrons scattered would not depend much on their energy. This is very different from the scattering of electrons from atoms or from a whole nucleus, where the "cross section" is strongly energy-dependent. The second prediction is something called "scaling." Bjørken convinced the experimentalists to plot their data not as counts versus energy, but rather as counts versus a new quantity called "the Bjørken-x variable," or just "x." "x" is a combination of energy-lost and momentum-lost information. When plotted, the data asymptotically reached the "Bjørken limit," which meant that the proton had subconstituents, contrary to Bjørken's own expectations!

Experimentalists were wary of such an abstract theory. An explanation in terms of current algebra may be a vindication for the current algebra model, but it left something missing in terms of physically understanding what was going on.

When Richard Feynman of Caltech visited the newly completed SLAC experimental facilities, he was intrigued by the data and by Bjørken's current algebra treatment and results. He is reported to have developed—in just a few days—something he called the parton model. The "parton model" is based on a field theory approach. (Field theory is what particle theorists who didn't do current algebra did.) The parton model essentially states that the proton is made up of pointlike subconstituents. These partons interact with themselves to a create a cloud of virtual partons around itself. This cloud is what gives rise to the spatial extent of the proton. Feynman was less than specific about the details of his partons. He did not define how many there are, or what charges they have, or if there is only one type or many types of partons. Still, he could accomplish two objectives with his model. He could reproduce the "Bjørken limit," and he could explain the new SLAC data in terms of the scattering of pointlike partons within the protons.

One of the refinements of the parton model was to realize that if the proton at high energies was seen as a collection of partons, then the scattering would be proportional to the sum of the square of the charges of all the partons in the proton. In fact, there are too many things going on in the proton for this to be a unique or distinguishing test. However, since most, but not all, of the same things are happening in the neutron, subtracting the scattering spectrum of the proton from that of the neutron provides a signature of the difference in their subconstituents. And that difference was the *fractional charge on one quark*, as predicted in the quark models of Zweig and Gell-Mann. This remarkable result is considered the most persuasive evidence supporting the quark theory, perhaps even raising its status from theory to recognized fact of nature.

So SLAC saw subconstituents of the proton and the neutron, which were called partons, the Bjørken limit was satisfied, and the particular parton model that rose to the top was the quark model. It had the right charges, the right spins, and the right isospins. But still no fractional charged particle was observed by itself, only the indirect signature! No free quarks.

As the number of searches for free quarks waned in the late 1970s, a new term appeared in the community—"confinement." This was at first merely a statement of the experimental observation. All quarks seemed to be confined in particles like neutrons and protons. Later, confinement came to mean that not only do quarks seem to prefer to be confined but, observationally speaking, they must be confined. Finally, confinement became linked to the dominant theory of strong interactions, that of the interactions between quarks, called quantum chromodynamics (QCD). Confinement is presently expected to be the most natural consequence of QCD—something that we will develop later.

Quantum chromodynamics is presently the guiding theory for physicists studying the world at a scale of 1 fermi (10^{-15} meter) or smaller. It is a complex theory whose complete solution has defied even the best mathematical physicists and the biggest computers. Yet it has in a few limited cases provided quantitative predictions confirmed by experiments. It has also prompted a class of models for a wide range of cases that are an approximation to this theory. Throughout this book a great deal of what I talk about will be described in terms of one of these approximate models, the "constituent quark model." At this scale we always need some type of model. From a model we can make predictions that can be confirmed in the laboratory—but going the other way is nearly impossible. A model also gives us pictures and descriptions of what is going on inside neutrons and protons, things we can talk about, discuss, and debate.

So, what is inside a proton or neutron?

- 3 -

The Players and the Stage

AFTER THE concertgoers settle down into their seats at Symphony Hall, while others are finding their places and stowing their coats and umbrellas, they are advised to glance at the program notes that the usher has left with them. Now, before the house lights dim, is the time to transform themselves from harried navigators of street traffic into placid receptors of the Muses. And the program notes can assist in this metamorphosis, by preparing the mind for that which is soon to come. The notes not only list compositions about to be heard, but also tell something about the world in which the piece was written. An attempt is made not only to draw us into the time and place of Mozart or Telemann, but also to draw us inside the composers' thoughts and methods—the orderly systems of Bach, the drama of Beethoven.

In the theater, we might feel that Willy Loman from *Death of a Salesman* is essentially no different from our neighbor. Perhaps his character is as familiar as an uncle we see every holiday. So we don't need a lot of preparation to meet him. Even Hamlet, that fated prince of Denmark, can step across nearly four centuries without introduction. We have often met his type in contemporary drama, and Hamlet himself is an old acquaintance whom we have known since high school English class. But, when we turn to the less familiar Shakespearean faces, the program notes are there to help us. At the very least they place Laertes as "Son of Polonius" and Polonius as "Lord Chamberlain." Then we are prepared, because when Polonius makes his entrance in front of Hamlet's uncle, it is not apparent to the uninformed what exactly his role in the court is. Shakespeare doesn't stop to introduce him—the torrent of verse nearly drowns the event, and there is no time to go back.

In the case of a Greek drama, we know that buried at its very heart are some universal themes, but still the need for good program notes is never stronger. Not only are all the names Greek to us, but the culture of the setting, the Hellenic references, and the style and structure of the play combine to make the drama a potentially challenging event. Although the culture of Antigone is at the most basic level truly universal, superficially it is a very different world from Manhattan or Montana. Even the choir that repeatedly parades out onto the stage to spout ballads of wisdom is sometimes more confusing than clarifying.

My point is simply this: when venturing into a new arena, program notes can sketch out the players or musical pieces for us, and explain what the stage in front of us is meant to be. Knowing something about the players and the stage doesn't detract from the performance—instead, it prepares us and allows us to be ready for the important events when they are played out in front of us. The program notes give us the static features. It is the performance itself that is dynamic and makes us want to care about the characters. So before we move into the dynamics of quarks inside the proton or the neutron, let's set the stage and introduce the players.

The stage is the nucleus of the atom. There is a factor of 100,000 between the height of a person and the diameter of a red blood cell (usually considered an "average cell"). There is another factor of 100,000 between the size of a cell and the size of an atom. Finally, there is another factor of 100,000 between the size of an atom and the world of the nucleus. Spanning fifteen orders of magnitude is not a conceptually easy task. I might make the analogy that it is the same as one penny out of ten trillion dollars, but I want to concentrate on an analogy whose dimensions are length.

Let us magnify an atom so it has truly macroscopic magnitude, increasing a carbon atom to the size of the Earth and a hydrogen atom to the size of the moon. Then a red blood cell would occupy the whole area inside the Earth's orbit around the Sun. If the cell was in the foot of an adult, the head of this magnified creature would reach to the next star.

What about the nucleus? With an atom magnified to the size of the Earth, the nucleus ranges from 50 to 250 meters across, roughly the size of a Broadway stage for hydrogen or deuterium to the size of a stadium for the heaviest elements. It is on this stage (or playing field) that the drama of quark dynamics will be played out. But before we peer into the neutron or the proton to see the quarks themselves, let us stop and look around. There is a lot to be seen on this scale, phenomena that contain a hint of the underlying quarks.

Long before the advent of the quark model we knew that the nucleus was made up of neutrons and protons. One other thing that we know about the nucleus is that the force between a neutron and a neutron is essentially the same as between a neutron and a proton, or between a proton and a proton. One of the first consequences of this symmetry is the use of the term "nucleon" to mean either a neutron or a proton. Not only is the force between neutrons and protons, or nucleons, symmetric, but they also have nearly identical masses. Therefore, a useful description of these two particles is that they are the same particle, with "something" changed. In nuclear physics that something is called isospin. As we will see later, it is the quark structure that lies at the heart of that similarity, as well as the differences.

Two other features of the nucleon-nucleon force, or "nuclear force," is that it is repulsive (positive) at about half a fermi, it is attractive (negative) at a fermi, and it essentially vanishes at about 3 to 4 fermis (see figure 3.1). That means that two nucleons will be oblivious to each other if they are separated by more than 4 fermis, and they cannot get closer to each other than half a fermi.

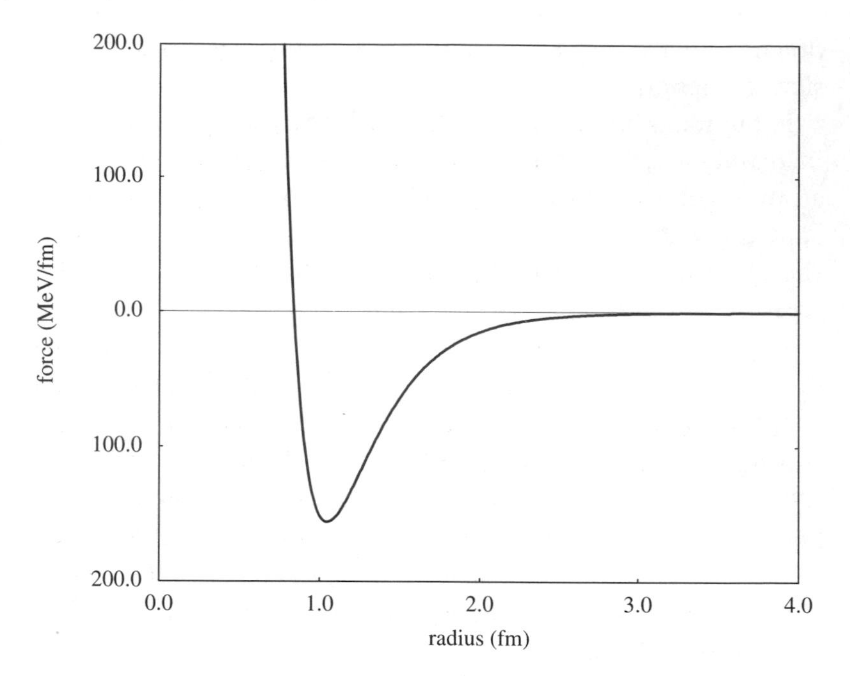

Figure 3.1 The nucleon-nucleon force is repulsive (positive) at about half a fermi, attractive (negative) at a fermi, and essentially vanishes at about 3 to 4 fermis.

The whole regime of nuclear physics is essentially defined by this distance scale. Since the scale of the quarks and the scale of the nucleon and nucleus are so similar, we should be able to explain nuclear physics, and in particular the nuclear force, in terms of the quarks. This will also serve as a good bridge between the physics of directly observed particles like protons and the physics of confined particles—quarks.

Now, before we submerge ourselves in the unique and particular causes of the nuclear force, let us revisit the more familiar gravitational force. With gravity, as well as with Coulomb or electrostatic forces, the force diminishes with distance, but *never* vanishes. A speck of dust does not have a great deal of gravity associated with it, and that force decreases as the square of the separation distance—according to Newton's law of gravity—but even at the other side of the universe that force is not zero. It may be "essentially," or "effectively," or "for all practical purposes" zero, but mathematically, it is still finite. This is because of the mass of the particle that carries the force—or "conveys the interaction." In the case of gravity the particle is called a graviton, an as-of-yet undetected particle (gravity is very weak). In electromagnetism the particle is the photon, a particle of light. Both of these particles are massless, and both of these forces therefore have an infinite range. The nucleon-nucleon force is not so simple.

There are several theories that describe the characteristics of the nuclear forces. The first successful theory was introduced by Hideki Yukawa, a Japanese physicist, who in 1935 postulated the existence of a new particle—the π-meson, or "pion" as we usually call it. He proposed that a nucleon could emit a particle that would carry away with it some momentum, and then be absorbed by a second nucleon. By this means momentum could be transferred between the nucleons, and the binding of the nucleons could be affected. That part of Yukawa's theory was no different from the exchange of photons in electromagnetism. What was unique about Yukawa's meson is that it had mass. If a nucleon emits a pion that has a mass, but the nucleon has lost no mass, then it has defied the law of the conservation of mass—one of the staunchest laws in physics! But there is an infinitesimal loophole in this law. According to Heisenberg's uncertainty relationship, there is always an intrinsic uncertainty in certain pairs of measurable quantities, such as position and momentum, time and energy, or distance traveled and mass. In some sense a system can "borrow" mass, if it doesn't reach beyond the region defined by the Heisenberg uncertainties. Given this constraint, and the average distance that the nuclear force reaches, Yukawa expected that the π-meson would have a mass of 100 to 200 MeV/c^2, or about a fifth of the mass of a nucleon.

At about this time—in 1937—cosmic ray studies first saw such a particle. Originally it was called a μ-meson. But, except for mass, it had all the wrong properties. In particular, it had spin $\frac{1}{2}$, just like an electron. Yukawa knew his particle had to have an integer spin in order to work as an intermediary between nucleons. Still, it was very tempting. It had the about the right mass, and its existence was at the time well documented.

In Japan, Yukawa and his colleagues were cut off from the rest of the scientific world due to World War II. So they formed an association they called the Meson Club, which met and presented papers and tried to hash out the problem of why the particle that Yukawa needed for his theory looked so different from the "μ-meson" that had been observed. Finally, in 1942, it was proposed that there were in fact two different mesons, the "π-meson" and the "μ-meson." This proposal resolved a great many problems, and later would be vindicated experimentally. In fact, by more recent taxonomy, we wouldn't even call the "μ-meson" a meson at all. It is electronlike, with spin $\frac{1}{2}$, and it usually has a charge of −1. That means that it is a "lepton" (in pseudo-Greek, a "light particle") that we now just call a muon. The π-meson, or pion, really is a meson (in pseudo-Greek a meson is a "middle particle"—a particle that goes in between other particles, or mediates a force). It has spin 0 and it can have a charge of +1, −1, or 0, which was just fine for Yukawa's needs.

After World War II, the Japanese theory was published worldwide. In the United States a similar theory had been developed by Robert Marshak and Hans Bethe—but Yukawa's theory was deemed to have been a great deal more developed and mature. Finally, in 1947, it was demonstrated that the μ-particle did not participate in the nuclear force, meaning that it couldn't be Yakawa's meson. Also in the same year cosmic ray experiments found another particle—Yukawa's pion! It had the right spin, charges, and even mass.

So on our stage of the nucleus we can view the nuclear force as a nucleon emitting a pion. The pion travels a fermi or so, and then is reabsorbed into a second nucleon. Physicists like to diagram such a process as shown below (figure 3.2). This is essentially a "Feynman diagram."

Feynman diagrams always sound mysterious, especially when a theorist pronounces that she will calculate (or occasionally has

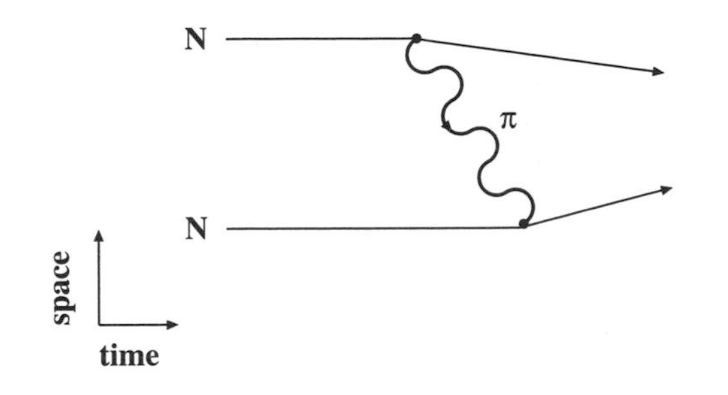

Figure 3.2 The Feynman diagram for a pion exchange.

already calculated) the occurrence probability of some process using a Feynman diagram. It is true that these diagrams are used as a type of shorthand notation to describe a long and precise calculation, but for us their primary role will be as a description of the sequence of events in an interaction. Actually, we can diagram any interaction—not just subatomic particles. Just bear in mind that each line represents a particle. The vertical distance shows separation (but not precisely) and the horizontal scale indicates time progression.

As an example, we can diagram a simple baseball sequence. A pitcher throws the ball, a batter hits it and runs. Pictorially we see this in figure 3.3. In a Feynman diagram we could represent this with figure 3.4. This tells us that the first event in the sequence was that the ball was thrown (A). The second event was the batter hitting the ball (B). We are also told that both the batter and the ball changed their motion as a result of the hit (C)—both back toward the pitcher—with the ball leading.

A more exciting baseball play is shown in figure 3.5. The sequence that we can read off from the diagram, and our interpretation since we have seen baseball, is something like this. (A) The ball is pitched. (B) The batter hits it. As a result of the hit, the batter runs and the ball flies. (C) As soon as light from the ball reaches the pitcher, the short-stop, and the left-fielder, they all start to run. (D) The pitcher and short-stop collide and stop their motion. (E) The left-fielder fields the ball, the runner sees this, turns, and heads for the dugout, and the team in the field heads

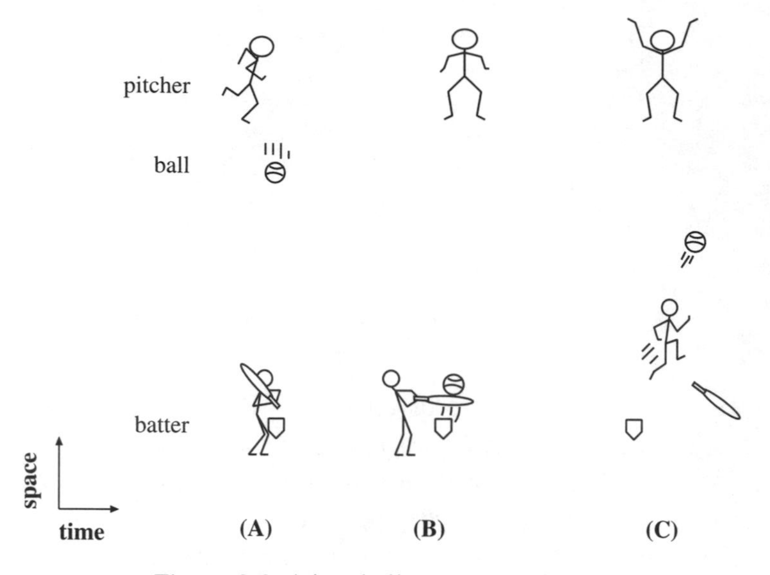

Figure 3.3 A baseball sequence of play.

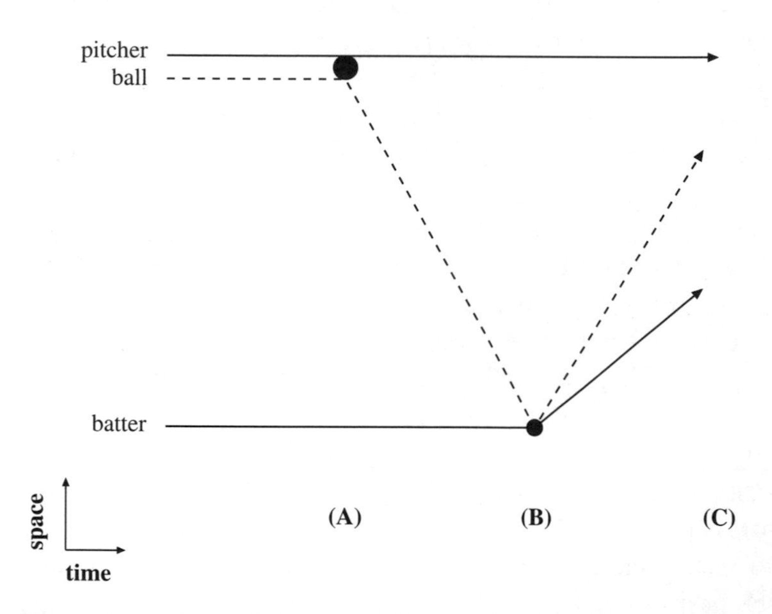

Figure 3.4 A baseball Feynman diagram for the same sequence as the previous figure.

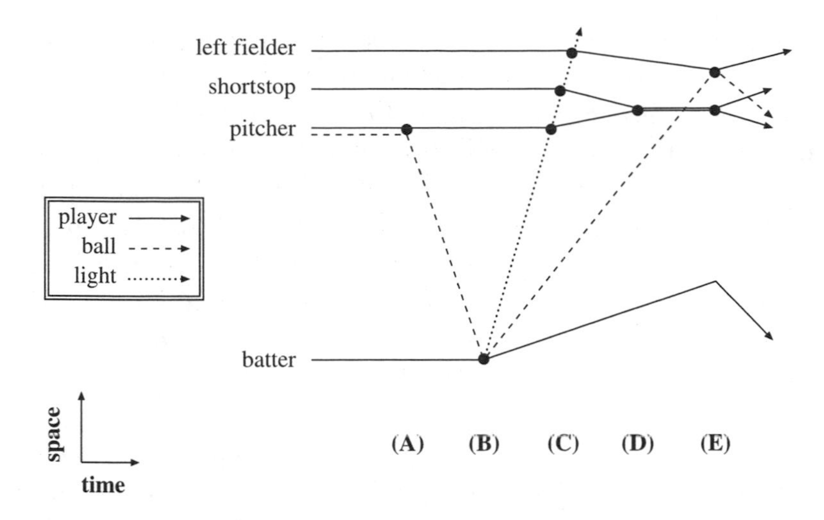

Figure 3.5 A baseball Feynman diagram for a more complex play.

to its starting positions. As we can see by this example, the diagram can keep track of the interactions, the time ordering, and the causes and effects.

Back to our "magnified stage," where the nucleons are playing out this drama. If the whole nucleus is the size of a playing field, then the proton and the neutron are balls 50 meters in diameter a few dozen meters apart, and there is a nearly continuous stream of pions, about 20 to 30 meters in diameter, being emitted by one nucleon and absorbed by another.

As I mentioned before, these pions have the peculiar nature of defying the conservation of energy in their role as intermediaries between nucleons. We therefore refer to them as virtual particles. Pions do not always have this type of existence. When we observe them in the laboratory—in the macroscopic world—they must conserve mass and energy, and then we refer to them as real particles. But in their role as the mediators of the nuclear force, what we observe is not a pion, but rather the fact that some protons and neutrons have bound themselves together to form the nucleus of an atom. The larger objects—the nucleus and the atom—do conserve energy and mass the way we think they should.

A problem that I always encounter when trying to visualize binding due to the exchange of virtual particles is the question: how does a pion know in which direction to travel? In the case of a ball rolling down a hill, the ball will follow the trajectory defined by the topography of the hill, following the lines of the gravitational force down. But the pion cannot perform the analogous motion, since the pions are the creators of the nuclear force. That is, the nuclear force can be viewed as the sum of the effects of innumerable pions being exchanged. One way to reconcile this problem is to realize that, given enough time, energy and momentum really must still be conserved. Our macroscopic perspectives are truly fundamental and well justified over larger distances or longer times. So if a pion were emitted in a direction in which it would not be absorbed, it would be lost to the nucleus, and mass-energy would not be conserved.

Yukawa's nuclear force theory, defining the nucleus in terms of neutrons, protons and pions, can describe the world on the scale of a few fermis (100 meters or so in our magnified world). If we increase our resolution a bit, these particles dissolve into vaporous clouds—now defined by the swirling collection of their subconstituents. With a diameter a third that of the neutron or proton (~ 0.3 fermi, or 15 meters on our stage), we see for the first time the "constituent quarks." In a very real sense the world of these quarks is just below the surface of the world of the nucleons, being separated by less than a full order of magnitude. There are few places in physics where the scales of two different classes of phenomena are so close. The difference between the nucleus and the atom is five orders of magnitude. The difference between the solar system and the galaxy is seven orders. Not only are nucleons and pions made up of quarks, but because of the similarity of scale we can expect that we should be able to describe the nucleus and the nuclear force in terms of quarks.

Inside neutrons and protons we see three quarks, each roughly a third of the size of a nucleon. Likewise, inside a pion we see two quarks—or more precisely, a quark and an antiquark. Now let us reconsider the pion exchange Feynman diagram for the

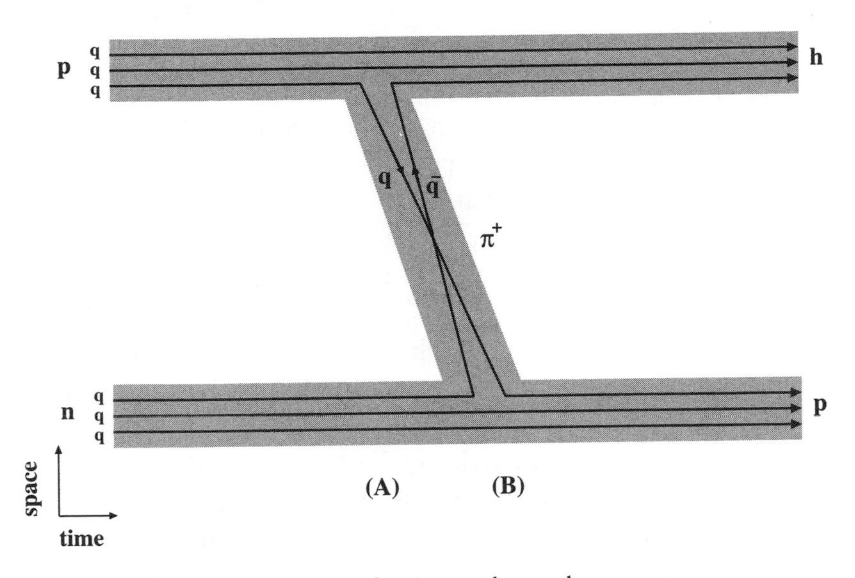

Figure 3.6 Feynman diagram of a pion exchange between two protons, in terms of quark exchange.

Yukawa model from a few pages ago. If we draw it in terms of its quark constituents, we will have a more complex diagram (see figure 3.6). Initially we started with a neutron and a proton—which is six quarks divided into two clusters. A pion is created, and within the pion a quark-antiquark pair. When the pion is absorbed, we lose this quark-antiquark pair and return to our initial complement of six quarks. However, momentum has been transferred and the neutron and the proton remain bounded.

The most peculiar feature of this diagram is the antiquark. Antiquarks are a type of antimatter—the stuff science fiction is made of—which really does have all those characteristics that drive futuristic spaceships. When a quark and an antiquark (of identical type) collide, they annihilate, giving off energy. What makes this real, and not science fiction, is that it is a common and everyday occurrence. At the time a pion is produced (time *A* in the above Feynman diagram), a quark-antiquark pair was created in the top proton. By creating a pair, everything that should be conserved, is—except energy. The charge on the antiquark is the exact opposite to that on the quark, so the charge created is zero. The spins and the "quark-type number" also add up to zero. Mass or energy is not conserved at this moment, but that is the same problem that we had with the creation of the pion. The solution is also the same: the system borrows some mass-energy and then returns it within a very short period of time. At time *B*, a quark-antiquark pair is annihilated and the pion is absorbed—giving up exactly the amount of mass-energy that was borrowed.

Another interpretation of what this diagram means, which is hinted at in the notation for the antiquark, is that an antiquark is the same as a real quark—only moving backward in time. In the case of all regular particles, the line in the diagram has an arrowhead pointing to the right to show that this particle is moving forward in time. But an antiparticle is marked with an arrowhead pointing to the left, as if it was moving toward the past. So we could interpret the Feynman diagram as saying that a quark from the bottom proton moves forward to time *B*. It then turns back in time to time *A* and then continues in the lower proton. In our first interpretation of this diagram, we borrowed mass in a small region of time/energy. In this second interpretation, we have borrowed time in the same small region—as allowed by the Heisenberg uncertainty relationship.

Do antiparticles really go backward in time? Descriptions of antiparticles as particles with all their qualities and quantum numbers backward, or as particles with the right qualities but moving backward in time, are absolutely indistinguishable. For example, think about a quark with a $+\frac{2}{3}$ charge. It would be attracted to an electron with charge of -1 by normal electrostatics—opposites attract. The antiquark of that same type would have a charge of $-\frac{2}{3}$, and be repulsed by the electron—like charges repel. However, if we view the antiquark as having the same quantities as the quark, but time-reversed, then it has the same $+\frac{2}{3}$ charge as the quark. We would expect it to be attracted to the electron, but instead of moving toward the electron we see the video being played backward, and the antiquark appears to move away from the electron. So the action of the antiquark is the same

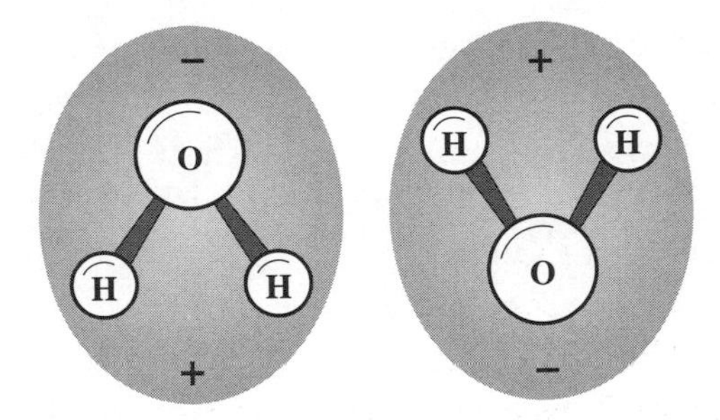

Figure 3.7 Water: Van der Waals force.

if it were a quark moving backward in time, or a quark with all of its quantum numbers backward.

The concepts related to descriptions of the nuclear force in terms of quarks—antiparticles, the violation of mass-energy conservation, and particles moving back in time—are all concepts that are consistent with Yukawa's theory of pion exchange. What really has changed is the scale of the players and the number of players on the stage. The last model of the nuclear force that I will describe carries this one step further, adding more details and working at an even smaller scale.

Originally, this last model viewed the nuclear force as a "color Van der Waals force," but more recently this has been modified into something called the "quark exchange" or "flip-flop" model. The term "Van der Waals force" is familiar in chemistry, where it is a very short-range force acting between two neutral molecules. For instance, the force between water molecules that gives rise to surface tension is Van der Waals. A water molecule—good old H_2O—is electrically neutral. That is, there are the same number of electrons as protons—ten each. Therefore, the whole molecule is neutral. However, if you are close enough to the molecule, there is some electrostatic force, for the electrons are concentrated at one end (on the oxygen atom), leaving the protons in the hydrogen somewhat bare (see figure 3.7). It is not as if the hydrogen is completely stripped, just that there is a tendency for the electrons to spend more time at the oxygen end of the molecule. This means that the molecule is polarized. If you are far away, the effect of the ten electrons and ten protons essentially cancels, but at a distance that is similar to the size of the water molecule, there is a charge. A second water molecule that is very close will line itself up such that its oxygen—slightly negative—will be near the first molecule's hydrogens—slightly positive.

This sort of short-range force seemed to be very similar to the nuclear force. The main difference is that in quark dynamics we have "color charge" instead of electric charge. In electric charge, there is one type of charge, which can be positive or negative, $-e$ or $+e$. In color charge, there are three types of charges, with the whimsical names "red," "green," and "blue." Each color charge can also come in positive and negative. We usually write them as $(r, \bar{r})$, $(g, \bar{g})$, $(b, \bar{b})$.

A proton or a neutron has three quarks. Each of the quarks is a different color—which adds up to a color-neutral, or color "white," particle. The rules that govern color interactions are (1) opposites attract, (2) the same color repels, and (3) white or neutral systems do neither. This is just the same as electric charge or even magnets. The difference is in identifying the opposite color. A red quark will repel a red quark and be attracted to an anti-red antiquark, but it will also be attracted to a green quark–blue quark pair, as is necessary to form the normal nucleons. This also seems to allow for the possibility of a color Van der Waals force.

So if two nucleons are widely separated, a quark in the first nucleon will view the three quarks in the second nucleon not as individuals, but rather as a collected set of quarks, whose combined color is white or neutral, and there is no attraction or repulsion. However, if the nucleons are close to each other, the story changes. Then a quark in the first nucleon can distinguish the quarks in the second nucleon. With our experience with water molecules lining themselves up head to tail, we can expect the nucleons to line themselves up in an orientation such that the red quark of the first nucleon is near the blue and green quarks and far from the red quark of the second nucleon (see figure 3.8). In this arrangement the primary lines of interaction

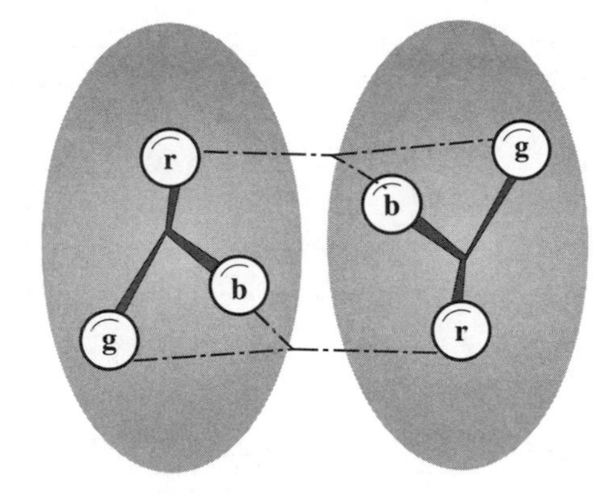

Figure 3.8 Color Van der Waals—quark flip-flop.

(indicated by the thicker lines in the figure) are still the ones that define the nucleon. But there could easily be secondary lines of force (indicated by the dashed lines in the figure) between quarks in different nucleons.

The problem with the color Van der Waals description of the nuclear force is that although Van der Waals and nuclear forces are short-range, Van der Waals is not short-range enough. A true Van der Waals force is very, very small as the separation grows, but it doesn't completely vanish. The nuclear force does vanish. However, in the pictures we have drawn lies the germ of the flip-flop model.

In the flip-flop model we recognize that what initially defines the nucleons are the connections indicated by the dark lines in this figure. But because they are so close to each other there is a probability that the quarks will re-pair themselves as shown in the Feynman diagram below (see figure 3.9). In this case the red quarks have been exchanged, or flipped, between nucleons, and that re-pairing is indicated by the lighter connecting line. This Feynman diagram looks essentially like two "pion-as-quark" exchange diagrams, one right after the other.

Back on our stage, a nucleon is three quarks racing around each other in a region 50 meters across. When these whirling objects approach each other, they can overlap. In the resulting confusion, two quarks could be exchanged, similar to a square

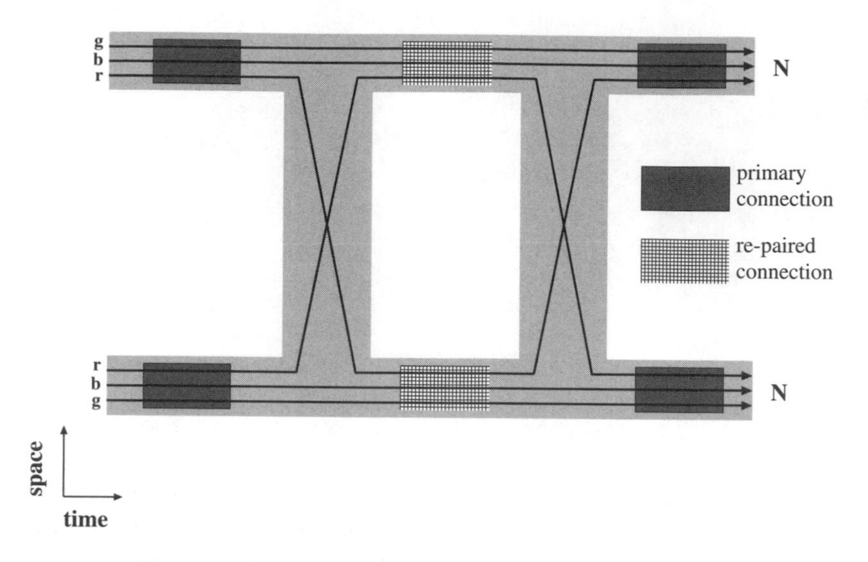

Figure 3.9 Feynman diagram for quark flip-flop.

dance, where everyone swaps partners in a way that works to keep the square of dancers together.

So we have three descriptions of the mechanism that underlies the nuclear force: Yukawa's pions, pions as quark exchange, and a quark flip-flop. Which is right? In one sense the flip-flop is embedded in the quark exchange, and the quark exchange is embedded in the pion exchange, so they all would seem to be equally true. But there are real distinctions related to scale and complexity.

The simplest model is Yukawa's. In his picture we had a neutron, a proton, and a pion—only three particles. Computationally this is the simplest. But this model breaks down as we approach the size of the nucleons themselves, when we must introduce the quarks as individual particles. In the quark exchange model we deal with six quarks, and then the creation and annihilation of a quark-antiquark pair. That is, we have up to eight particles and we can describe matter down to the distance between quarks. The third model—the flip-flop—has six quarks, but also two sets of interquark binding as well as multiple permutations of rearrangements. This last model should work down to a scale defined by the interquark binding.

So as we go to models with higher resolution we also gain degrees of complexity. But that is just a computational problem that we should be able to overcome with faster and better computers. Why then do we still maintain the pion exchange model? It is because nature really does produce a particle called the pion. It has been observed and measured. Physicists can even form them into a beam that can be steered across a laboratory and used as an experimental probe. In fact, if we had only looked at the nuclear force we might never have guessed at the existence of the quark. There are small details about the nuclear force that are not completely accounted for by the pion. However, with the addition of other, less abundant particles, such as the ρ-meson or the ω-meson, which also have been observed, a complete description of the nuclear force does arise. If I want to describe nature down to a scale of one fermi, I don't need quarks. All the effects of quarks and antiquarks, and their interactions, generally sum up to very discrete and unique bundles—the nucleons and the mesons.

So why bring up the quark picture at all? The quark model still performs the same function among these exchange mesons as it did among the particles that in the 1950s and 1960s drove Gell-Mann to the eight-fold way. The pions (π^+, π^0, π^-), the ρ-mesons (ρ^+, ρ^0, ρ^-),the ω-mesons, and so on are all explained in terms of two types of quarks. And finally, quarks do come into their own at the right scale, at less than a fermi.

One final comment when discussing quarks and their scale. What is the size of a quark? I said that a constituent quark was roughly a third of a fermi across. But a constituent quark is not a true or "bare" quark; rather, it is a bare quark wrapped up in self-interactions. Theorists believe that a bare quark is pointlike, and all experiments confirm this. In reality, experiments can only say that the quark is smaller than the resolution of the largest accelerator. With the accelerator at FermiLab, near Chicago, producing a trillion eV of energy, we can say that a quark is less than one-thousandth of the size of a nucleon. In the magnified analogy with which we started this chapter—with a human reaching the stars, atoms the size of Earth, and nucleons fitting inside a playing field or on a stage—we can say that a bare quark must be smaller than 10 centimeters, or about $2\frac{1}{2}$ inches, across.

- 4 -

The Nature of the Evidence

IN THE HISTORY of the study of matter, what sets quarks apart from every other level of matter is that we are not able to isolate and observe them directly. Not only is this limitation inconvenient and frustrating for the investigator, but it has also created a psychological barrier. The quark hypothesis has been with us for thirty years and we know a great deal about what quarks are like, yet there are still physicists who treat them as merely a "mathematical convenience."

Throughout the history of physics we have been able to confirm our hypotheses about the basic constituents of matter by distilling and isolating those constituents. In the eighteenth and nineteenth centuries, chemists were busy isolating the "elements" before Dmitri Mendeleev could compile his *Periodic Table of the Elements.* A hundred years ago J. J. Thomson, at Cavendish Laboratories in Cambridge, England, launched our study of the atom and the fundamental particles by observing the electron in detail. This was soon followed by Ernest Rutherford's discovery of the nucleus and later the proton, and shortly thereafter James Chadwick's observation of the neutron. But we have never isolated a quark. In accelerators we have created fireballs with 1,000 times more energy than is necessary to create a proton. Yet we still have never seen a quark, and there are many reasons to think that we never will.

Yet this synopsis of the study of matter is not only simplistic but deceptive. A truer history should highlight the linkage between size and what counts as direct evidence. The trend in the study of matter is not just a migration to smaller and smaller scales, it is also a migration to more abstract evidence, evidence that can only be understood through a growing reliance upon theoretical interpretation.

In this light we can look at the work of Antoine Laurent Lavoisier, who isolated oxygen during the late part of the eighteenth century. Although he was dealing with a substance that he could not directly "see," he knew that it occupied space and displaced water. More important, flames burned in its presence and were extinguished in its absence. In some sense, he knew that he had oxygen because he could put it in a bottle, label it, and put it on a shelf.

Likewise, J. J. Thomson, in 1897, is generally credited with the "discovery" of the electron—or at least the determination of many of its characteristics. Initially he was trying to understand what cathode rays are. At the time there were two competing hypotheses about their nature. One school of thought maintained that cathode rays were "aether" or "wavelike," while the other school proposed a "particlelike" explanation. Thomson's apparatus, the cathode ray tube (CRT), is essentially the same as a television tube. What he and his contemporaries called a cathode ray we would now identify as a beam of electrons. He subjected this beam to electric and magnetic fields and measured how the beam was deflected. He also focused the beam on a metal receptacle, an anode, and measured how much charge, heat, and energy were deposited. From these measurements he could infer the velocity of the electron, its charge to mass ratio, and the sign of its charge. Furthermore, he knew that the mass of the electron was 2,000 times smaller than the mass of a hydrogen atom, and finally that cathode rays were made out of particles—the electrons.

At this stage, with the introduction of the first elementary particle, already the nature of the evidence has changed. For Lavoisier, matter had classical traits: it occupied space and it had mass—it could be weighed on a scale. J. J. Thomson's electron was a very different creature. He never attempted to measure its size—it would have been a futile endeavor, as it still looks point-like even to the most sensitive experiments we have today. His measurement of mass, unlike Lavoisier's, was not gravitational, but related to energy; the kinetic energy converted to heat, or the inertial resistance to being deflected. Finally, he never "saw" the electron. Rather, he saw a bright spot on the phosphorescent coating of his cathode ray tube.

In a series of experiments at the University of Manchester in England, from 1910 to 1918, Ernest Rutherford and his assistant Hans Geiger established the existence of the nucleus, and later the proton. They scattered alpha particles, emitted from a radioactive radium source, off a thin gold leaf. Rutherford knew from his earlier work at McGill University in Montreal that alphas are relatively massive and so he expected the alphas to pass through the gold with little or no deflection. He set up detectors—both a Geiger counter and a zinc sulfide screen (more sensitive than Thomson's phosphorus screen)—on the far side of the gold from the alpha source. He observed a large number of alpha particles punching through the gold leaf. This is what he expected, for the prevailing model held that the gold atom, or any atom, was a thin, homogeneous cloud of positive charge mixed with electrons, which couldn't stop an energetic alpha. Rutherford then moved his detectors to the other side of the gold foil and still observed alphas. These were alphas that had scattered *backward* off some dense and massive core—the nucleus of the atom.

These experiments are precursors to our modern study of quarks. Not only do they introduce us to the nucleons, but they also point to the value of scattering experiments and the development of new detector technologies. Scattering experiments and the direct descendants of the Geiger counter are at the core of any modern high-energy or nuclear physics laboratory.

Later, in 1919, Rutherford was able to scatter alphas off nitrogen gas and liberate a hydrogen nucleus—a proton. He then demonstrated that the nuclei of all atoms contain protons and even went on to estimate the size of the proton. He measured it to be 7×10^{-13} centimeter, similar to a modern measurement and by far the smallest thing measured at the time. However, Rutherford realized that something was missing—and he postulated the existence of the neutron to explain unaccounted mass and even isotopes.

It fell to James Chadwick, at Cavendish, to detect the neutron, the elusive uncharged counterpart of the proton. Chadwick's neutron detector consisted of essentially a flask of hydrogen in front of a proton detector. Neutrons are neutral and will not flash on a zinc sulfide screen or produce ionization in a Geiger counter. However they can collide with and knock out protons from hydrogen. These newly liberated protons can then be detected in a standard proton detector. In one sense Chadwick's detector was but a slight improvement on the methods of Rutherford and Geiger, but in another sense his detector had a whole new element, the hydrogen in front of the proton detector.

So Chadwick had detected the particle that couldn't be detected. He had done this by completely understanding the way in which protons would interact in his device. He knew that the protons that his Geiger counter saw had originated in the hydrogen, and that they had not been liberated by alphas or protons. What he was left with was a neutral particle with about the same mass as a proton.

This is the dominant trend in the detection of more and more elusive particles: more sensitive detectors depend on a more detailed understanding of how the detector works and what other phenomena it might see. The physicists' observation of a reaction or event under study becomes a longer and longer chain:

$$\alpha \rightarrow (\text{neutron}) \rightarrow \begin{pmatrix} \text{proton} \\ \text{from} \\ \text{hydrogen} \end{pmatrix} \rightarrow \begin{pmatrix} \text{ionization} \\ \text{in Geiger} \\ \text{counter} \end{pmatrix} \rightarrow \begin{pmatrix} \text{electrical} \\ \text{impulse} \\ \text{\& meter} \end{pmatrix} \rightarrow \text{physicist}$$

The working physicist realizes that observations are not just limited by the weakest link of the chain, but are even weaker. The uncertainty in the final results is an accumulation of the uncertainties from each link. The uncertainty also depends on how much data are available, how large the statistical sample is. So in a modern experiment we have replaced the observation of a flash or the clicks of a Geiger counter with an array of detector systems, many of which have well over 99 percent efficiency, as well as specialized data collection computers that can gather data from thousands of detectors, millions of times each second. Yet, before we discuss the modern descendants of the fluorescing screen and Geiger counter, we need to understand what we might detect, if it is not the quarks themselves.

The identification and measurement of a particle, be it an atom, a proton, or a quark, is only one kind of measurement—and only the first stage of an investigation. The goal of physics is to understand how the world is put together, and not just what it is made of. It is the dynamics of a system that gives rise to many of its characteristics, characteristics that can be experimentally measured. If quarks were building blocks that were simply added together to form protons and neutrons, atoms and molecules, in the way Legos snap together, it wouldn't matter if they were ultramicroscopic billiard balls, or just "convenient mathematical concepts." But they don't just add up. A proton is not just the sum of three quarks—it is also a product of the way the quarks dynamically interact with each other.

When John Dalton was proposing his atomic theory of chemistry (1803), far more important to him than a complete list of the elements were observations, such as the "law of constant proportions." This law stated that molecules are made up of definite ratios of the elements. For example, the ratio of hydrogen to oxygen in water is *always* two to one by volume, and eight to one by weight. Dalton recognized that this would follow if the world was atomic in nature. If an elementary substance is made out of discrete atoms, it can only combine in discrete units. He proposed that the unique and defining feature of the atoms of an element was the number of "hooks" and "eyes" that it had on its surface. His sketches of atoms look a lot like burdock, those seed pods that attach themselves to your socks when you walk through an autumn meadow. These pictorial "hooks" of Dalton end up corresponding to the valence electrons of modern chemistry.

In atomic physics, if we had only observed the electrons and the nucleons, we would have missed one of the most startling revelations of the century. It was the measurement and explanations of the atomic spectrum that really opened up the world of quantum mechanics. It was those bright lines in the familiar spectrum that inspired Niels Bohr to postulate the privileged orbits of the electron about the nucleus. "Privileged," because the electrons did not radiate away their energy and spiral into the nucleus in the way a "classical," non-quantum mechanical system would. Erwin Schrödinger explained why these orbits were privileged by "quantizing" the atom, and explained the relative intensity of each line as a result of the probability and the overlap of "wavefunctions." Werner Heisenberg explained the width of the spectral lines as an intrinsic property of wave mechanics—usually expressed as an uncertainty relationship. Finally, P.A.M. Dirac gave us a detailed relativistic quantum mechanics, which included the spin of the electrons and the microscopic splitting of the spectral lines.

All of these new developments in theoretical physics—the Schrödinger equation, the Heisenberg uncertainty relationship, the Dirac equation—were first postulated to explain the atomic spectrum. The spectrum, in its more abstract and generalized form, continues to be the single most important way of presenting both theoretical and experimental results today. For a model or a theory to be tested, it must be able to calculate some type of spectrum. For an experiment to be a useful test, one must be able to take the raw data and present it as a spectrum. It is the meeting point of these two branches of investigation, theory and experiment.

The quintessential atomic spectrum is that of hydrogen, with a bright red line at one end and a blue line at the other, with a faint blue and even fainter violet lines beyond that. The intensity of the red line, for example, means that a large number of "red" photons, with wavelength of 6,562 Å, or 1.89 eV of energy

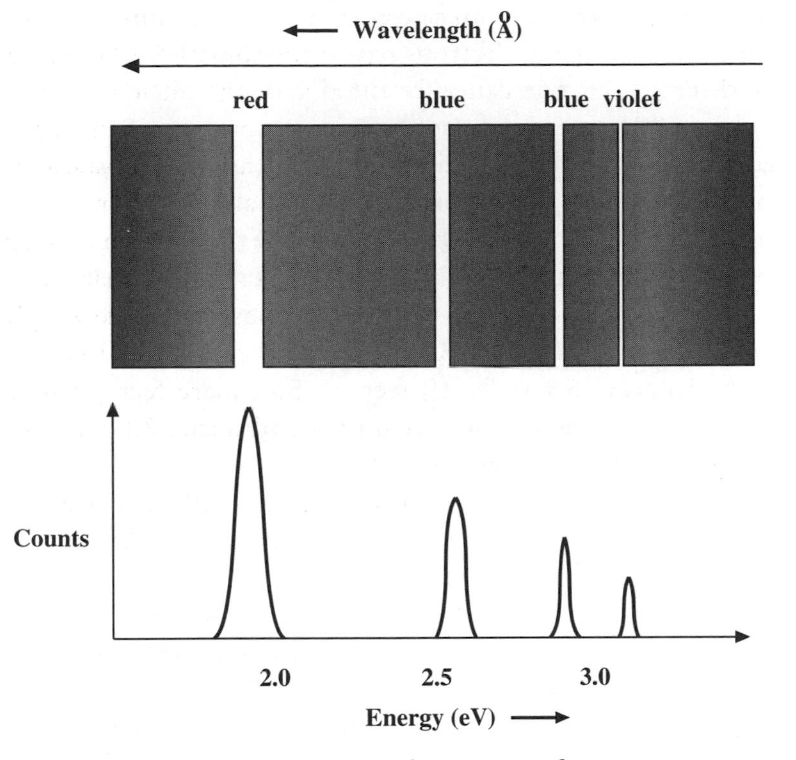

Figure 4.1 Spectrum as histograms of counts.

(color, wavelength, and energy are all equivalent) were detected—either by a photographic plate or perhaps by eye. So we could present a spectrum as a histogram of the number of photons detected versus the energy of each photon (see figure 4.1).

With a model such as the Bohr model of the atom, or better yet the quantum mechanical atomic theory, these lines are explained as energy released when an electron drops from a higher energy level (or orbit) to a lower level. What we measure is the number of photons as a function of energy. What we infer through our models and theories is the energy bound up in each orbit.

In nuclear and high-energy physics we measure something called a cross section, which is essentially a generalized or abstract spectrum. As an example, a simple cross-section measurement would be to direct a beam of electrons at a thin film of carbon, and count how many electrons pass through without being scattered. From this one can infer the size of the nucleus. But the probe and the target need not be microscopic to measure a cross section. Imagine that you are trying to measure an obstacle in a dark hallway. Simply station your friend at the far end of the hallway, then take 100 baseballs and throw them down this dark and mysterious corridor. When your partner reports that seventy balls arrived at his end you know that the obstacle blocks roughly 30 percent of the hallway. By measuring the cross-sectional area of the hallway (5 feet by 10 feet, or 50 square feet), you can calculate the cross-sectional area of the obstacle: 30 percent of 50 square feet—or 15 square feet.

So a cross section is a measurement of an effective area. If a proton were a solid little ball, and we could fire infinitesimally small BBs at it, we could simply measure its area and size. But we don't have infinitely small BBs. The best we can do is an electron. Although an electron is as small as one could hope for, it doesn't collide with the proton in the same way a BB would. It interacts via the electrostatic, or Coulomb, force. That means it can scatter off the proton even if it just passes near it and doesn't actually touch it. In fact, this is not so bad, since we understand Coulomb scattering so well. Also, from the viewpoint of the electron, what defines the boundaries of the proton is where the proton's charge is located. So a proton can be "spongy," and not just a hard sphere. If the electron has a lot of energy, if it is thrown hard, it can penetrate deeply. If it loses no energy, it misses. If it loses a lot of energy, it hits near dead center.

We can pick up one more piece of information: at what angle did the electron scatter? In billiards the energy lost and the angle scattered are proportional, a result of both the *conservation of energy* and the *conservation of momentum*. But that is not purely true for our experiments. Energy can be conserved by using some of it to deform the nucleus or the nucleon. They act more like rubber balls than billiard balls, as they can hold energy by being "stretched" or "compressed" or "excited," and then release that energy a moment later. And this too is a useful tool for understanding the interior of a nucleon.

So with these two pieces of information, energy lost and angle scattered, we can infer how deep the electron penetrated and how much charge the electron encountered. Also, we can probe the various modes or deformations by which the nucleon can temporarily store energy. This is the type of information the Stanford/SLAC group (Hofstadter and later Kendall, Taylor, and Friedman) had when they measured the size of the proton and "discovered" the quark. The interpretation of this cross section—or spectrum—as a function of angle can be quite complex. The general shape is dominated by the well-understood Coulomb scattering, but there are numerous bumps and valleys, places where the energy is right for the nucleus or nucleons to absorb some energy into a distortion or to resonate.

But there is potentially more information available than just the energy and scattering angle of the electron. Quite often some other particle or photon is knocked out of the target as well. We can measure its energy and angle too. For example, if we are scattering an electron off a carbon target, we will sometimes also knock out a proton. In this case we can construct a "five-dimensional cross section," a spectrum which is a function of five different measurements: two energies and the three angles involving the electron beam, scattered electron, and ejected proton.

The number of possible measurements increases as we add complications. What if we knocked out two protons? Or perhaps a pion and a neutron? But the problems with "coincidence" measurements are twofold. First, they are hard to perform. If the detection of one particle is difficult, then the detection of two coincident particles is difficulty squared. Second, sometimes there is no new information. The second or third particle may only confirm the conservation of energy and momentum.

The measurement of a cross section will always be a difficult experiment. So before an experimenter is ready to invest years of labor from dozens of physicists, and thousands of dollars' worth of beam time, he will look for guidance from a theorist. The theoretical physicist, or theorist, may be able to point out a part of a cross section that is uniquely interesting, sensitive to intriguing physics.

When experimentalists resolve to "measure the most important cross section in the nucleon-quark transition region," that is, study matter at the scale where quark descriptions and nucleon descriptions are both approximately valid, it is a daunting task. Still, plans are drafted, laboratories are built, and tasks are organized. The work load is subdivided in a cooperative manner such that a single measurement might draw upon the equipment and experience of a collaboration of 100 physicists and an army of technical support.

The task of the theorist is equally daunting: start from the best-established theory—QCD—and try to predict the world we see, or at least the cross sections measured by the experimentalists. But this is a task that defies organization and subdivision. The best techniques are not obvious, and QCD is a difficult theory to work with. Much as quarks insist on sticking together, the equations of QCD do not readily unravel. And although it is our favorite theory and has passed every challenge to date, QCD is not the only theory out there.

Theorists have approached this problem from a number of sides. Some concentrate on only one particular aspect of the theory, reducing it to a simplicity that they can tackle. But these simplifications may no longer describe the real world. These papers will be peppered with phrases such as "consider the case of a one-dimensional universe," or "in the case of a massive quark . . . ," or "with only an isotropic distribution. . . ." One hopes that techniques might be developed, or perhaps a trend will be recognized, which will point in the right direction for a future, more realistic calculation to pursue. An alternative approach is to create a phenomenological model. Here, many of the subtleties of a full theory might be averaged over, or replaced with experimental results. The constituent quark model, which guides a great deal of the description in this book, falls into this category. We have been speaking of this model as if it were well established and universally recognized. Yet two theorists starting from the same model will differ in their results. One might indulge in an approximation early in her calculation that allows a simple interpretation of results later. The second theorist may choose less approximation, perhaps substituting computer simulations or numerical solutions, and end up with a result that is more accurate, but harder to understand and less instructive.

This whole noncentralized approach may seem a bit chaotic, but it lends itself well to the mind-set of the theorist. Theorists tend to work in groups of rarely more than three. Too many theorists, and there would never be an agreement on techniques and appropriate approximations.

To demonstrate some of the considerations that go into a calculation, let us consider describing a Δ particle from a theorist's point of view. We will see some of the rudimentary parts of the constituent quark model, the role of conservation laws, quantum mechanics, relativity, and spin. We can finally pull all these together to create a cross section.

In QCD, and all quark models, the proton and the neutron are made up of three quarks. Two of them have their spins aligned (*parallel*) and one is *antiparallel*. The Δ particle is essentially the same as these nucleons, except that it has the spins of all three of its quarks aligned. But what is spin?

Spin is a property that all particles have which is meant to conjure up visions of a child's top, or the spin a pitcher can put on a curve ball, or the "English" that can be put on a cue ball as it skids across the green felt of a billiard table. In all these cases, including quarks, the best way to "see" the spin is to observe what happens when the ball or particle collides with something. When a billiard ball hits the cushion at the edge of the table, it is not simply reflected; how it reacts depends upon the direction of its spin. If the spin rolls it along the cushion, it bites into the cushion and bounces away at a shallower angle, closer to the cushion than before the collision. If a backspin skids it along the cushion, it bounces off at a larger angle. Sometimes we can use this type of scattering to measure the spin of a particle.

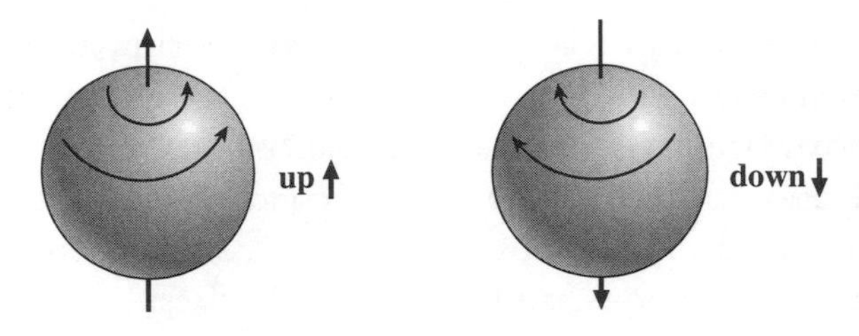

Figure 4.2 Spin up and spin down.

Figure 4.3 Combination of two spin $\frac{1}{2}$ quarks can be +1, 0, 0, or −1.

But quarks and electrons cannot have spin in exactly the same way as a billiard ball. They are pointlike objects, spheres with zero radius. How can the equator spin around the axis when the equator doesn't go around the axis, but is in the same place as the axis? What they share with a billiard ball is the way their "bounce" depends on their spin. More important, they share the same type of mathematics. For instance, a billiard ball spinning has essentially two orientations; "up," which is counterclockwise when viewed from the top, or "down," which is clockwise. The spin axis may be tipped, but when it comes to bouncing off the cushion it is only the part of the spin that is up or down which counts (see figure 4.2).

What makes spin interesting is when two particles collide or combine. There are no "cushions" in a nucleon—only other particles with spin. When two billiard balls are on a table—or two particles are bound together—there are four possible spin combinations (see figure 4.3). In the first combination, the sum is twice the spin of one ball. In the second and third cases, the spins cancel and sum to zero. In the last case, the magnitude of the sum is the same as the first case—but with the opposite orientation.

In the case of quarks and electrons, the magnitude of their "intrinsic" spin is quantized into steps of $\frac{1}{2}$. This is where the analogy to billiard balls breaks down. Intrinsic spin is something unique to the quantum world. If a ball rotates faster, its "spin," or angular momentum, increases. If it stops rotating, its angular momentum vanishes. But a quark *always* has spin $\frac{1}{2}$.

If we have a pair of quarks, we can add up their spins, like the billiard balls pictured above, to +1, 0, or −1. If we have a triplet of quarks, like in neutrons and protons, the spins can add up to $\pm\frac{3}{2}$ or $\pm\frac{1}{2}$. Our familiar nucleons, the protons and the neutrons, are the spin $\frac{1}{2}$ combination. Protons and neutrons have two quarks with their spins aligned, and one quark with its spin in the opposite direction. The difference between the $+\frac{1}{2}$ and the $-\frac{1}{2}$ is just a matter of orientation. What about the combination of quarks in which all three are aligned with each other? This combination—with spin $\frac{3}{2}$ —is called the Δ particle.

The Δ was first observed at the Chicago Cyclotron in 1952, a dozen years before the quark hypothesis was put forth. Although its discovery predates Gell-Mann and Zweig, it is still instructive to start with a quark model of the Δ, to try to calculate what an experimenter would observe.

In most ways a Δ is the same as a nucleon. It has three quarks that orbit each other in a roughly symmetric manner. The notable exception to their similarities is the fact that the quarks in the Δ are all aligned. The other major difference is their mass. The Δ has a mass of 1,232 MeV/c^2 whereas the nucleon has a mass of 937 MeV/c^2—a 30 percent difference. In fact, this mass difference, and all the other unique characteristics we will encounter, are a result of the spin flip of a single quark.

Buried in the equations that govern the dynamics of any quantum mechanical particle with spin is a term referred to as a spin-spin interaction term. In the case of atomic physics, the interaction between the spin of an electron and the spin of a nucleon will alter the binding of the electron by one part in a million. Tiny as that is, it can be seen in careful spectroscopic studies as the "hyperfine-splitting" of a line in the spectrum into sublines.

When we move on to the scale of the constituent quarks, this hyperfine-splitting has grown to a dominant role, a mass splitting of 30 percent!

It may seem peculiar that "interactions" can modify masses so dramatically. In our everyday experience the mass of a composite object will have the same mass as the sum of its parts, if we include the glue as a part. And it most certainly cannot have less mass than the sum of the parts. But in the subatomic world, masses do work in this peculiar way. The deuteron does weigh less than the sum of the masses of its constituents: the neutron and the proton. This is because when we talk about masses in the subatomic world, we mean the energy required to create such a particle or combination of particles. When a proton and a neutron are combined to make a deuteron, excess energy is released. Thus, it takes less energy to create a deuteron than a free proton and neutron, and therefore the deuteron has less mass. Always when we measure mass we do it through some dynamic property. We collide the particle into a detector and measure its kinetic energy. We bend its trajectory in a field and measure its momentum. It is always from these types of measurements that we infer mass. The point is, mass is energy (remember Einstein).

In addition to mass, our quark model can tell us about the dynamics of the Δ. When a collection of three quarks, a nucleon, is excited, the spin of one of the quarks can flip to form a Δ. It takes a lot of energy to form that Δ (the 300 MeV/c^2 mass increase) and there is little to stop it from dumping that energy and decaying back into a nucleon. Therefore, the lifetime of the Δ is *very* short: 5×10^{-24} second—or 5 yactoseconds. This is roughly the amount of time it takes light to travel 1 to 2 fermis, or to cross two protons. Even if you hit a Δ out with an electron from the world's largest accelerator and effectively increased its lifetime due to relativistic time dilation, it still would not be able to travel all the way to your detector!

What we can detect are the decay products. When the Δ decays back to a nucleon, it will release that 300 MeV of energy. This could be a powerful photon, a gamma ray. (In contrast, an X ray

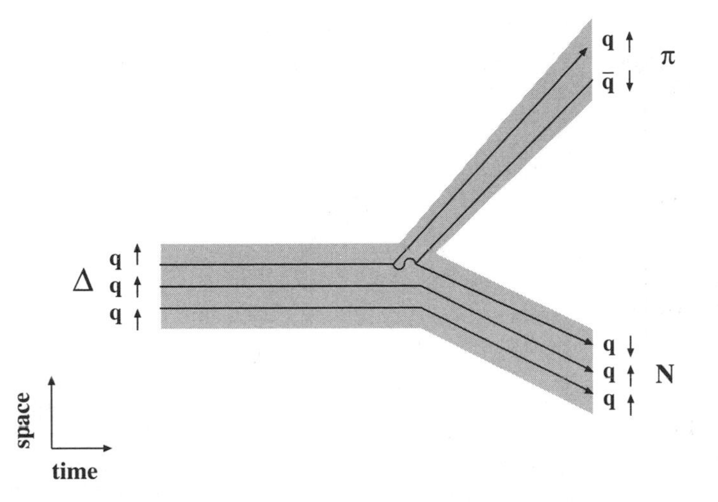

Figure 4.4 The decay of the Δ particle, as seen on the quark level.

will have dozens of Kilo electron Volts [KeV] of energy, four orders of magnitude weaker.) Alternatively, the energy may be used to create a pion, which needs only 140 MeV, and some motion.

In fact, a quark model based calculation will predict that the latter is by far the dominant decay mode. Experimentally it is verified that it is the decay of choice 99.4 percent of the time. The Feynman diagram for this decay is shown in figure 4.4.

This creation of a pion is much like the process that created the nuclear force pion of chapter 3, except that it is not made out of borrowed energy for a fleeting moment of time. This is a real pion, which will last 2.6×10^{-8} second—26 nanoseconds. In the world of the nucleon, that is practically forever. A pion will travel a few meters before decaying. That is galactic distances compared to quarks and nucleons, and these pions will even show up in our detectors. A peak in the energy spectrum of the pion is a good signature of a Δ.

One last characteristic of the Δ is that it doesn't have an absolute and unique mass. Rather, it has a mass of 1,232 ± 120 MeV/c^2. This is a consequence of its short lifetime and the Heisenberg uncertainty relationship. If the lifetime of a particle

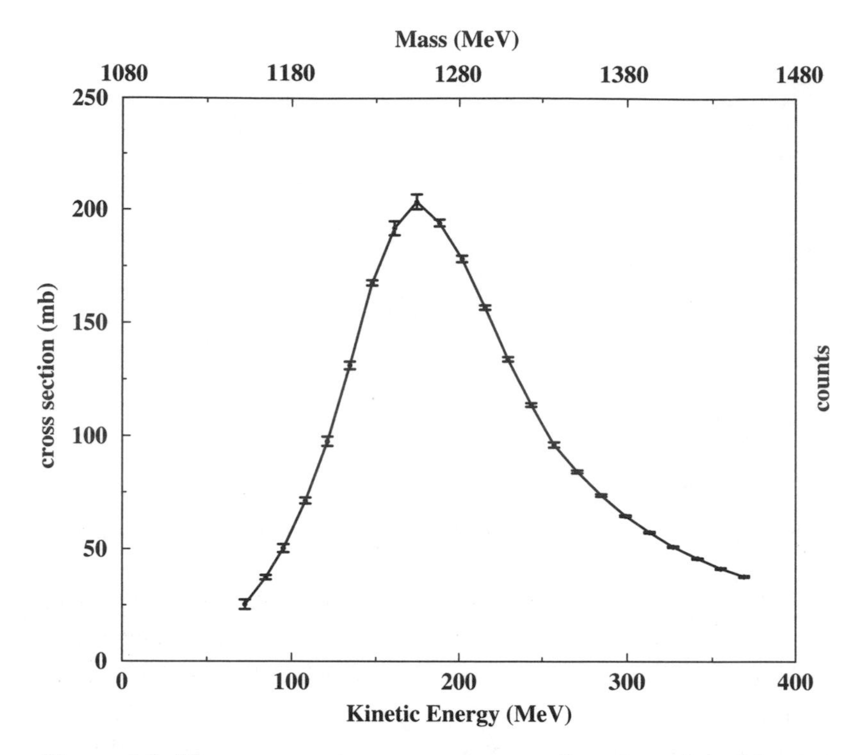

Figure 4.5 The cross section, or counts, as a function of the kinetic energy of the proton and pion, or the mass of the Δ (kinetic energy + $mass_p$ + $mass_\pi$ = $mass_\Delta$).

is short, then the uncertainty of its energy or mass is large. A short lifetime and fast decay lead to a broad spectral line. Conceptually the Δ has not had enough time to settle into a well-defined and stable configuration before it decays. Therefore, the mass of the Δ can vary by as much as 10 percent.

So when a theoretician thinks about what a Δ looks like, she builds her ideas out of a whole range of theories and models and principles. She has a constituent quark model: three quarks each with spin $\frac{1}{2}$. The Δ is the spin $\frac{3}{2}$ combination of the quarks. The spin-spin interaction causes a hyperfine-splitting, a difference in the mass between the nucleon and the Δ of 30 percent. The decay releases enough energy to create a pion. And finally, due to the short lifetime, there is a spread of range of masses.

These are the aspects of the Δ which a theorist considers in her thoughts and calculations, but when it comes time for the experimentalist to look for a Δ, we must present the models in terms of a spectrum or cross section. In figure 4.5 is shown the theorist's curve, cross section-versus-Δ mass, and the experimentalist's raw data, counts-versus-pion energy. The experimentalist will count pions and measure their energy. But he knows that the pion's energy is exactly dependent on the mass of the particle that decayed. He also knows how to convert his raw counts to cross section by correcting for detection efficiency and the number of electrons, in analogy to correcting for the number of baseballs that were thrown at the target.

So, finally, it is here in the cross section that theory and experiment meet, and the fleeting Δ rises up in the data to be counted.

- 5 -

Measuring a Rainbow

ONE OF THE great ironies of the study of nature is that to see anything very large (the edge of the universe) or very small (quarks) we need truly enormous instruments. On a clear night under the stars, you can see forever, all the way to the shoreline of the cosmos—to the first instant of time—but you wouldn't recognize it. Likewise, this book contains about 20 octillion (20×10^{27}) quarks, right now in your hand, but by staring at paper you really don't "see" quarks. The light with which you see this book—and everything else—has a wavelength of 5,000 to 7,000 Å (1 Å = 10^{-10}m ~ 1 atom), which means that each light wave will span a few thousand atoms. That sets the limit of what we can see with our eyes, even with the aid of lenses and optical microscopes. To see something the size of an atom we need a source that will produce light with a 1Å wavelength. That corresponds to a photon with about 2,000 eV of energy, which is an X ray. A photon with 200,000,000 eV or 200 MeV of energy will just barely resolve a nucleon. An additional order of magnitude will allow us to start to probe the inside of a nucleon on the scale of the constituent quarks, and this is where the massive accelerators come into play. The accelerators are effectively light sources, and the detectors are the eyes that see the nucleons and the effects of the quarks.

Perhaps an exhaustive description of the experimental measurement of a cross section is a major digression from the central emphasis of this book—a description of quarks in nucleons. But by presenting the efforts of experimentalists in detail, I hope to instill some sense of where these images come from and to what degree we have confidence in them. We are in the early days of subnucleon structure studies, and undoubtedly some of the images in this book will be modified with time. But most of them are built upon the meticulous labors of thousands of experimentalists worldwide. Through their works we have an impressive array of data to support the theories and models with which we make our pictures of quarks.

What do the "data" look like from an experiment? When the data are collected they are a stream of digital bits on a wire plugged into a computer. Alternatively, when the data come in they are electrical pulses related to a discharging wire, or tracks left by a particle as it coils up in the magnetic field of a detector. Or they are a spectrum, the sum of a million events—counts versus energy and angle. When the data really come in, they are a curve, a cross section on a page of a journal.

Clearly the "data" are all of these things, much as a book is ink and paper, letters, words, sentences, images described, and ideas conveyed. From another point of view, a book is the thread that connects authors, editors, artists, publishers, printers, booksellers, librarians, and readers. Finally, a book is only a collection of quarks until its message is internalized by the readers, matured in their minds, and hopefully added to their perspective, enlightenment, and enjoyment.

An experimental study has at least as many aspects. In one sense an experiment is born as an idea long before data are actually taken. It takes years to gather a collaboration of physicists, develop targets, design detectors, and of course find the funding for all this activity. After the data are collected, it still takes time and energy to analyze and synthesize those data. Finally, theses, dissertations, reports, and papers are written and published.

But that is only one facet of an experiment. From a laboratory director's point of view, there are accelerators, detectors, and targets. There are computer networks, data taking, archiving, and computer analyses. There is even the logistical problem of getting a thousand people and a similar number of machines to work together. There is accounting, personnel, grounds, utilities, safety, and so on.

From a personal perspective, involvement in this kind of research is a lifetime commitment. It starts with that decision in high school to take that extra math course; then majoring in physics in college; later graduate work, post-doctoral research, a research scientist position, or perhaps a faculty position at a university. It can be a cradle to grave obsession. It some respects it may sound like indentured servitude, but research as a lifelong career has a major redeeming quality: the problems that were asked yesterday are different from the problems of today, and tomorrow promises new and unique challenges.

One final facet of this kind of experimental study is that questions about quarks and nucleons span hundreds of experiments as well as careers. They even cross laboratory boundaries and oceans. The problems discussed in these pages and the visions of quarks presented here are derived from dozens of laboratories worldwide, hundreds of experiments, thousands of people, and billions of dollars of expenditure.

To describe all this is beyond the means of a single book, perhaps even beyond the scope of most libraries. So I will do what all experimentalists do—I will dip my net into this seething sea of activity and pull out a sample which will, hopefully, be representative.

There are laboratories around the world that contribute to our understanding of the arrangement of quarks inside nucleons. Each laboratory is a custom-built facility, with its own one-of-a-kind accelerator, handcrafted detectors, and uniquely tailored experimental program. The MIT-Bates Laboratory near Boston has built a beam-ring that will cause the electrons in the beam to pass through the target repeatedly—billions of times—until they finally scatter. The "Synchrophasotron" in Dubna, near Moscow, can accelerate ions of almost any kind in what the *Guinness Book of World Records* has recognized as the world's heaviest machine. In laboratories from Sweden to Canada to Japan there are accelerators with unique traits: from energies to intensities to purity of beam. They have beams of every imaginable type: electrons, neutrons, protons, ions, and even high-intensity photons.

The laboratory I will describe in great detail in this chapter is the Thomas Jefferson National Accelerator Facility, in Newport News, Virginia. Until 1996 it was called CEBAF (Continuous Electron Beam Accelerator Facility). Jefferson Lab doesn't have the largest, or the highest-energy, accelerator, but it is the newest laboratory with some of the highest-quality beam and most sophisticated detectors in the world.

Jefferson Lab was born out of a 1979 report by the Nuclear Science Advisory Committee (NSAC), which identified the need for a new accelerator and complement of detectors to study nature at the scale of the nucleon-quark transition. One of the most important aspects of the laboratory's organization is that it is an institution built out of inclusion and collaboration. To start with, the grounds, the buildings, and the accelerator are operated for the Department of Energy (DoE) by the Southeastern Universities Research Association (SURA). As its name implies, SURA is a group of universities from the southeast United States that wanted the next national laboratory to be built in their part of the country. They joined together and proposed Jefferson Lab to fulfill the mission identified by NSAC. But that was just the start of the cooperative and inclusive nature of the laboratory. When it came time to build the detectors, some of the construction and most of the management emanated from Jefferson Lab, but the design and fabrication of the detector components took place at over a hundred institutions, from over forty countries worldwide. Although it is a U.S. government laboratory, millions of dollars have been spent by the Italians, the French, the Russians, and others.

The laboratory itself occupies a half-square-mile parcel of land in a high-tech park in Newport News, Virginia. The nerve center of the campus is CEBAF Center. At first it may appear no different from any other office building in the high-tech park, except that in the lobby a monitor continuously displays the status of the accelerator: what energy and how much beam is being delivered to which experiment. Also the lobby is papered with posters from local schoolchildren who have participated in one of the many educational outreach programs of the laboratory. Its auditorium is as likely to play host to a conference of physicists preparing experiments as it is to a general public lecture on a whole range of science topics, or even the occasional piano concert.

Other buildings on the grounds house technical libraries, fabrication and testing laboratories, and the offices of teams of physicists and engineers. Nestled next to a wood lot is the "residence facility," on-site housing for visiting scientists. This is a wonderful resource when your experiment is running around the clock: a bed is only a ten-minute walk away. In fact, with computer terminals and Internet connections in the rooms, physicists can keep an eye on their experiments even when they are off-site and trying to get some rest. But it is back through the white oak and pitch pine grove from the residence facility, past the white-tailed deer and the testing laboratory, where the whole reason for the laboratory is focused: the accelerator and the experimental areas.

The chain-link fence that surrounds the accelerator is primarily for safety rather than security. The guards at the gate are not interested in protecting secrets or "security clearances"; rather, they will insist on checking your safety clearance—have you been trained to work around radiation (a minimal hazard at this laboratory), high voltage, or, most hazardous, cryogenics such as super cold liquid helium. Inside the fence the earth seems alive, although most of the noises are from ventilator fans. There are a number of buildings: the accelerator control building, the refrigeration plant, the "counting house" where the experiments and the detectors are controlled, and three great domed mounds. The accelerator itself is buried 10 meters underground in a racetrack-shaped tunnel 7/8 of a mile around. The three mounds cover and shield the three cavernous experimental halls where the accelerator finally dumps its electron beam into targets and detectors.

But experiments do not happen on their own. Experiments are born out of questions. Questions are raised by previous experiments with unexplained results—or by theories that seem sound but predict radically different results than other models. Questions are raised in journal papers, or presentations at colloquia, or conference sessions. Experiments are often conceived in the back row of an auditorium, triggered by a comment of the speaker. Sometimes they are born in conversations between colleagues during a conference, or sharing an experimental shift. Sometimes it is while browsing the latest journals in the library, or in the quiet morning or soft evening hours when the world is still and thoughts can mature. Whatever the origin, it is often a beautiful moment.

Yet experiments are not born full grown and mature. An experiment must go through a long gestation period before the basic ideas are ready to face the world. The originator will want to first know if it is a practical experiment before going public with his ideas. Usually, through a series of calculations or computer simulations, he will address such questions as: Can you really see the sought after effect? Is the experimental apparatus sensitive enough? What other effects might imitate or mask what you are looking for? There is also a series of less scientific questions such as: Is there enough interest in the physics community to support this measurement? When compared to other experiments, is it really unique? Are there enough human and financial resources?

A well-developed experimental proposal is more than a description of what the experiment can achieve. It is also a detailed study of the limitations of the experiment. The experiment may have a signature, such as the pion-nucleon energy peak shown at the end of the previous chapter. It will also always have physical and experimental limitations.

Physical limitations are concerned with how the signature compares to "background." Background means an event that could be mistaken for part of the signature—like trying to recognize the music over the static on a distant radio station. For example, when seeking the Δ, one might detect a proton and a pion which didn't originate from a Δ: perhaps they have no common history and only coincidentally have energy similar to the peak of the Δ resonance. The proton simply may have been knocked out of the target, the pion may have resulted from a completely separate interaction of the beam with the target. Worse yet, the original beam may have accidentally scraped the edge of the beam pipe, or the frame holding the target, or something else. One of the particles could have come from outside the experiment. Some backgrounds we can minimize with a good design, but other backgrounds are intrinsic to the experiment. Often the signature is small compared to the well-understood background. In that case we can "subtract" the background to obtain our signature. Occasionally that might mean a preliminary experiment to first measure the background (in a different region than the signature), which could spur a separate line of experiments.

The apparatus will have intrinsic limitations. If a detector can measure the energy of a 1 GeV proton to a resolution of 10 MeV, and the signature is only 2 MeV wide, it won't be seen clearly. The experimenter may hunt around for a laboratory that has a detector with higher resolution. If one doesn't exist, ideally he could build a new one, but that is expensive (many millions of dollars), and the experimental program would have to be unique and critical to justify it.

Finally, there is a "statistical" limitation with every kind of measurement. If an experimenter saw 100 events in one hour, he would expect to see 100 in the next hour, plus or minus ten. The size of the sample tells how well things are known, in this case to 10 percent. The more data one collects, the more confidence one has in the measurement. To increase the amount of data, he could build a thicker target, but this will distort the beam and make the initial energy of the event uncertain. The experimenter could also increase the intensity of the beam, but accelerators have upper limits, and they usually run nearly flat out. Finally, the experiment could run for a longer time, but beam time is expensive and the competition for beam time is intense.

Once an experimentalist has convinced himself that his experiment is feasible, he will build a collaboration. The experiments are far too complex and expensive for a single institution to tackle. Usually he will present his ideas and estimations to other experimentalists, generally those who work in the laboratory where the experiment will take place, or to physicists who have a unique interest in that particular kind of experiment. He will ask this group for comments, criticism, and perhaps collaboration.

Help from collaborators comes in all forms. Most institutions have developed a specialty, perhaps a unique target or an auxiliary detector or a special analysis computer code. Besides resources, collaborators can help fine-tune the experimental proposal with additional information on resolutions or backgrounds or new data for comparison from other laboratories. When the collaboration feels an experiment is feasible and compelling, it is time to approach an accelerator laboratory to request beam time. If an experiment requires a lot of new equipment, it may also be time to go to a funding agency to seek a grant. In the United States the primary agency for funding this kind of laboratory and research is the DoE. The National Science Foundation (NSF) is also an important player.

Beam time requests are handled slightly differently at every laboratory, but in general, they are reviewed by a Program Advisory Committee (PAC), who advises the laboratory director as to whether an experiment is both feasible and interesting. The collaboration spokesperson, usually the physicist who originally conceived the experiment, will present the case for the experiment to the PAC. This is often done in two stages. First, a long and detailed written document is submitted. Second, a presentation is made, in which the spokesperson must present the experimental motivation and design. In this presentation, the spokesperson must field any conceivable question and defend the experiment against all criticism. This is a *very* serious endeavor. Beam time is not only expensive, but it is also a finite commodity. A laboratory at its best can produce only 365 days of beam a year, yet every group of experimentalists would like to have as much beam time as they can. More time means more data, better statistics, and a better measurement.

About half of all proposals are turned down. Perhaps the experiment is being done elsewhere, or it is technically too difficult at the time, or the PAC has doubts whether the measurement can really answer any questions. Another fraction of the proposals may be "contingently approved," that is, approved if a technical problem with a target is resolved, or if the expected results from another laboratory confirm the existence of the peak that is to be studied, or if a detector resolution improves. Finally, some proposals really are approved; beam time is allotted and eventually even scheduled.

With an approved experiment in hand, there can still be several years of preparation. There is always the reengineering of targets and detectors for this particular laboratory and experiment. There is the training of young students and the testing of equipment before it is moved into the experimental area of the laboratory.

The final preparations include an endless list of interconnecting systems to check and monitor. There are the computers that measure and control the target positioning and pressures. There are the thousands of elements of the detector. There are the dozens of physicists who will fly in from around the world to contribute their unique skills, and also to provide the general staffing for around-the-clock data-taking shifts. There are also the new students who will spend months and years away from their schools, working at the laboratory. Not only do they need to learn the physics and the laboratory, but they need to find apartments and transportation. There are a thousand other details that require the attention of the experimentalist, but it is not a solo venture—there is an experienced collaboration behind every experiment. Still, as with any expedition into the unknown, it is by dealing with the details that disasters can be averted and the probability of success enhanced.

On the day when an experiment finally goes on-line, it is as complex as any space launch. The event should be accompanied with all the drama of the Kennedy Space Center, but it is not quite as heart-stopping. A thousand things can go wrong, but most of them can be fixed with only the loss of precious beam time, precious data. What if the power supplies to the magnets trip, or short? What if there is a gas leak? Or what if the data are out of sequence and scrambled? Or what if there are no data at all? All of these things have happened, but with adequate planning and experience, generally the experiment, after some false starts, will be ready to collect its valuable data.

"Hello—Machine Control Center. This is the crew chief. Could we have beam? 4 GeV at 1 microamp please? Thank you." Of course, the accelerator operators have known what would be requested for months—the experimentalist had sent them a detailed run plan—but now it is time for the accelerator to deliver.

At one end of the accelerator (see figure 5.1), far from the experimental area, is the electron source: a hot, glowing wire—like a lightbulb. In fact, the simplest source, the thermionic electron gun, is essentially the same as the back of a television tube. In the gun, electrons are boiled off a wire, a cathode (remember J. J. Thomson's cathode ray electron experiment?) by passing a current through the wire. The electrons are drawn electrostatically away from the cathode by the positive charge on an anode. There is a hole in the middle of the anode through which some of the electrons pass to start the beam. In between the cathode and the anode is a 100,000 volt potential drop, so the energy gained by an electron is 100 KeV.

Some experiments will require a very different source, if they need electrons with their spins lined up in one particular direction. In that case, instead of a hot wire, a laser is focused on a gallium-arsenide (GaAs) crystal. Because of the crystal structure

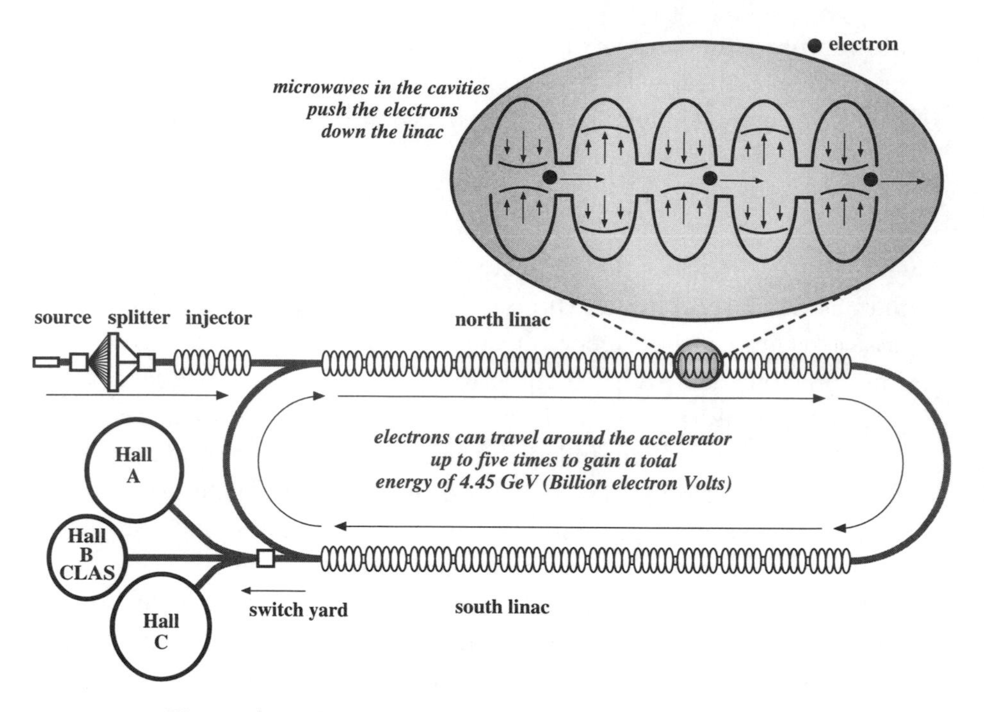

Figure 5.1 The accelerator at the Thomas Jefferson National Accelerator Facility.

and the polarization of the laser, the electrons that are dislodged tend to have their spins aligned. It was pointed out to me by one of the physicists working on the polarized source that the laser that is used is so weak its beam could be broken by his hand without any ill effects on his hand—but it would shut down the entire accelerator. At the other end of the accelerator, the electrons have gained a few billion eV of energy, and you do not want to put your hand anywhere near it!

So far the electron has 100 KeV of energy and has traveled only a few inches on its trip of about 4.5 miles. Next the beam is "chopped" into micropulses, which are easier for the accelerator to push. This is done by sweeping the beam over a metal plate with 3 holes in it. Sweeping an electron beam by rapidly changing electric fields is not hard. Television sets sweep beams across the screen about 15,000 times a second. An accelerator will sweep its beam about 800 million times a second, to produce 2.5 billion pulses (remember, there are 3 holes.) The rest of the injector shapes each pulse to be about 1 millimeter long and gives them a total of 40 MeV of energy.

All the dimensions of the accelerator are set by the chopper. Two and a half billion pulses means that they are separated by 400 picoseconds, and since they have approached the speed of light they are spaced about 12 centimeters apart. Therefore, 12 centimeters is the length of the accelerator cavities.

The main accelerator is in a concrete tunnel shaped like a racetrack. It is 7/8 of a mile long and buried 10 meters underground. In each straightaway is a linac, or linear accelerator. The linac is made up of a long series of hollow cavities, each of which is the size and shape of a slightly squashed grapefruit, with the electron beam passing through the centers of all of them. Microwaves are pumped into each cavity, where the waves oscillate back and forth. The trick to accelerating the beam is to get the beam pulse to enter the cavity just as the microwaves start to move forward, so they push the electrons. With cavity after cavity pushing the electrons, they end up effectively riding a wave down the linac. One of the side effects of the oscillating microwaves is that they tend to oscillate the electrons in the metal walls of the cavity too. Normally this would lead to a great deal of heating, but at Jefferson Laboratory the whole accelerator is cooled in liquid helium. At these temperatures, 4°K above absolute zero, the cavity walls become superconductors and the heating and energy loss from the electrons sloshing in the metal vanishes.

After one time around the accelerator, the beam has gained an additional 800 to 1,200 MeV of energy. It can now be used for an experiment, or else sent through the accelerator again to gain an additional 800 to 1,200 MeV. This can be repeated up to five times. In the fifth and final pass, the electrons have about 100 times the energy they had after the injector, but in both cases their speed is just below the speed of light. In fact, after passing through one linac, a 6 GeV electron will have only gained 5 picoseconds, or 6 milimeters, over a low-energy 40 MeV electron.

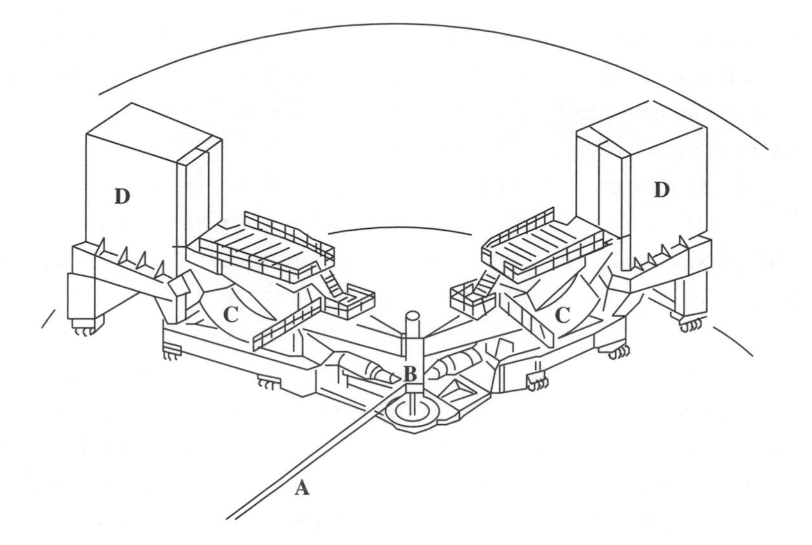

Figure 5.2 The two High Resolution Spectrometers in Hall A. (A) beam pipe, (B) scattering chamber, (C) main magnet, (D) detector hut.

This means that they can both ride the same wave down the accelerator. Isn't relativity amazing!

Finally, after the electrons have been accelerated to the desired energy, the switchyard will send them toward their final destination: one of the three experimental halls. The first hall, Hall A, contains two monstrous spectrometers (see figure 5.2). Each one is over four stories tall and 25 meters long. They always remind me of ships with their bows moored to a post in the center of the experimental hall. Decks, stairs, and railings add to the nautical image.

Coming out of the wall, 4 meters up in the air, is a shining stainless steel beam pipe, 3 inches in diameter, which is the only part of the accelerator an experimentalist normally sees. The electrons come out of the accelerator barrel down that beam pipe and into the scattering chamber, which is a cylindrical box on top of the central post. Inside the chamber is the target, often a thin sheet of carbon, or a waterfall for oxygen and hydrogen studies, or sometimes a glass vial of gas such as helium.

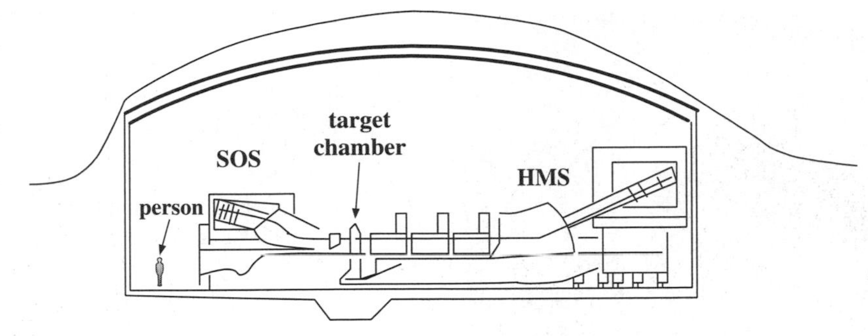

Figure 5.3 The electron's view of entering Hall C, with the Short Orbit Spectrometer on the left and the High Momentum Spectrometer on the right.

Some of the electrons are scattered off the target, possibly to reach a detector. However, most do not react but pass straight through the hall, where they are absorbed into the beam dump. Of those which are scattered, only a small fraction head toward a spectrometer. In the spectrometer is a magnet the size of a city bus. This magnet is the heart of the spectrometer. It will bend the trajectory of a low-momentum particle more than that of a high-momentum particle, so that particles of different momenta are physically spread out by the time they reach the detector hut.

The detector hut is a garage-size, heavily shielded building balanced on top of a spectrometer. Inside it, various detection elements can measure when the particle arrived and how much the magnet bent its path. A precise measurement of this bend is also a precise measurement of the particle's energy and momentum. The Hall A spectrometers can measure momentum to an accuracy of one part in four thousand, a truly world-class measurement.

In Hall C there are also two spectrometers, the High Momentum Spectrometer and the Short Orbit Spectrometer (see figure 5.3). These were the first two spectrometers to come on-line at Jefferson Lab in 1995. As the names indicate, the first one is designed for high-momentum studies. In fact, when the energy of the electron beam was increased beyond the 4 GeV Jefferson

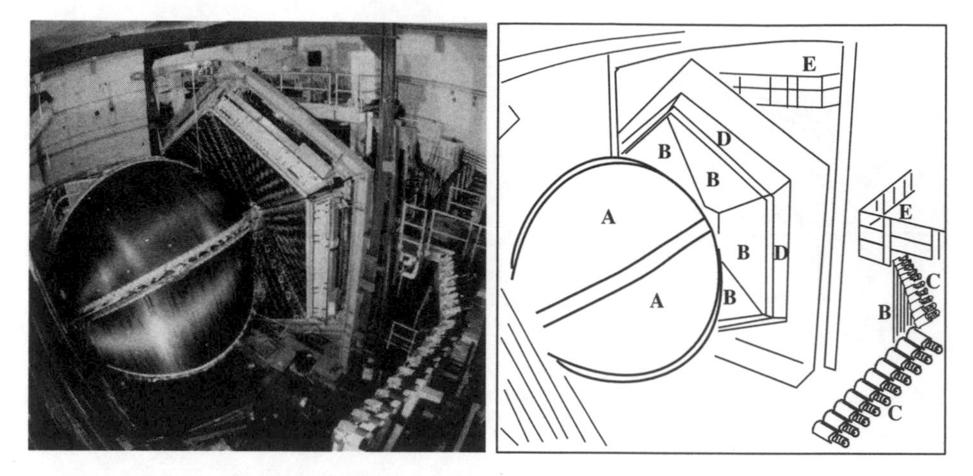

Figure 5.4 The Hall B CLAS detector unfolded. (A) outer wire chamber—6 meters in diameter, (B) time-of-flight scintillators, (C) photo multiplier tubes (PMTs), (D) calorimeter, (E) catwalks—the detector is four stories tall. (Courtesy of Thomas Jefferson National Accelerator Facility [Jefferson Lab])

Lab was originally designed for, the High Momentum Spectrometer was ready and waiting.

The Short Orbit Spectrometer is designed with a curious combination of magnets that twist the trajectory of a particle into an *S* pattern. By doing this it loses some resolution, but the detector hut is only 10 meters from the target, less than half the distance of the three other spectrometers. It was designed this way to study particles with short lifetimes, such as strange mesons like the lambda, which would decay before traveling all the way through a long spectrometer.

The spectrometers in these two halls can all be pivoted to various angles around the scattering chamber. Hall B, the last experimental hall to come on-line, is radically different. Hall B contains CLAS, the CEBAF Large Acceptance Spectrometer (see figure 5.4). (The collaboration that built CLAS chose not to change its name when the laboratory changed from CEBAF to Jefferson Lab.) CLAS is a 4π detector. The designation 4π for this kind of detector arises because a sphere of radius one has a surface area of 4π, and this detector is like a sphere, collecting, detecting, and measuring particles that come out of the target in any direction.

CLAS has the same components as most other major detectors: magnets, wire chambers, Čerenkov counters, timing scintillators, and calorimeters. But when you are trying to catch particles in all directions, the geometry becomes quite different, starting with the magnets. Since we want all particles traveling in any direction to go through the magnetic field, the field must be wrapped around the beam pipe. To accomplish this, six huge coils are arranged around the target, radially from the beam pipe. The coils generate a toroidal or doughnut-shaped field. Electrons that come down the beam pipe and don't scatter will pass through the hole in the doughnut-shaped field and into the beam dump. Those electrons that are scattered, as well as protons, pions, and other particles that are knocked out of the target, will travel through the magnetic field and have their trajectories bent. The high-momentum particles are only slightly deflected, the low-momentum particles are dramatically deflected.

It is not enough to scatter electrons and knock out protons; we must also detect the particles and extract the information from their distorted trajectories. The detector components in CLAS are packaged into six wedges, shaped like giant pieces of an orange, and inserted in between the magnets. When completely assembled, CLAS becomes a sphere that encompasses the target. Each wedge contains a number of layers of detector elements that can measure the trajectory, momentum, and energy of the particle. The first three layers that the particle encounters are the wire chambers, or drift chambers. Their purpose is to measure points along the particle's trajectory. By locating the trajectory in chambers at roughly 1, 2, and 3 meters from the target—before, in the middle of, and after the magnetic field—we can determine where the particle originated, which way it was going, and what its momentum was (the bend in the field tells us this) (see figure 5.5).

A wire chamber is essentially the same as the device designed by Hans Geiger for Ernest Rutherford in the 1910s. A charged

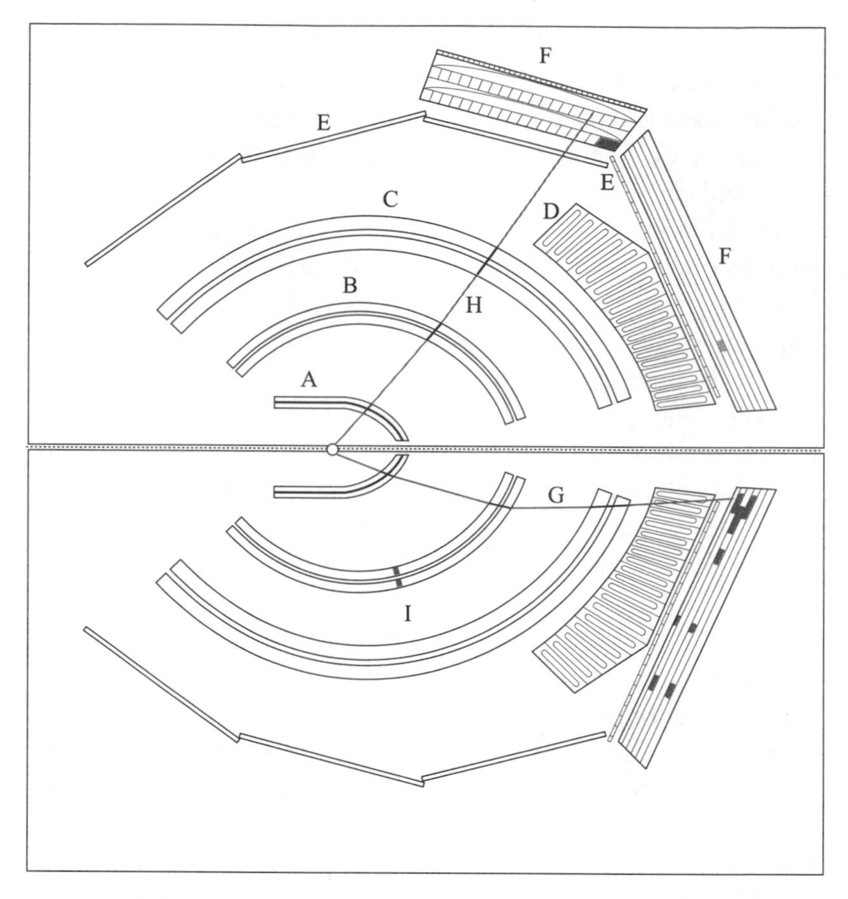

Figure 5.5 The cross section of CLAS with a reconstruction of an event. (A) inner wire chamber, (B) middle wire chamber, (C) outer wire chamber, (D) Čerenkov counter (E) time-of-flight scintillators (ToF), (F) calorimeters, (G) track of an electron, (H) track of a proton, (I) track of an unidentified particle.

particle passing through a gas will leave a trail of ions, atoms that have had electrons knocked off them. The ions, which are positively charged, will drift toward a negatively charged sense wire. The charge of the ions is then deposited on the sense wire which, after amplification, is seen as an electrical pulse in our computer, or as a "click" in a classical Geiger counter. The first major improvement over Geiger's original tube is that we have thousands of sense wires (34,000 in CLAS). Therefore, by knowing which wire collected the ions, we can get a rough idea of where the particle went—to within a centimeter. The second improvement comes from timing the pulse. By knowing when the particle goes by, and how fast the ions drift, we can calculate how close the particle passes to the sense wire—to about a tenth of a millimeter. With this information, CLAS can measure momentum to about half of a percent.

The next layer of CLAS is the Čerenkov counter. Čerenkov light is produced when a particle that is traveling at nearly the speed of light in air or vacuum enters a material that is optically "thicker," where the speed of light is much slower. The particle must slow down to comply with this new speed limit. To slow down, the particle must dump energy, which we see as Čerenkov light, named after the Russian physicist who first identified the process. In an experiment at Jefferson Lab, the electrons will all be traveling at nearly the speed of light, whereas the pions and nucleons will be much slower because they are much more massive. By selecting a material with the right optical density, the electrons will all emit Čerenkov light, whereas the other particles will pass through, invisible. Since these experiments are initiated by an electron from the beam, identifying the electron can often be key to unraveling and deciphering the data read out of CLAS.

The next layer out in this onionlike detector is the time-of-flight (ToF) scintillators. This layer is about 3 meters from the target and consists of over 300 "bars." Each bar is made of scintillating plastic, 2 inches by 6 inches, from 0.5 meter to nearly 4 meters long and weighing up to several hundred pounds. The over 300 bars fit together to form an almost complete shell around the inner detectors. As described in chapter 2, scintillating plastic is clear and looks like Plexiglas, but scintillates—produces a metallic blue light—when a particle passes through it. At each end of the scintillator bar is a photo multiplier tube (PMT). A PMT is a highly sensitive light detector that can sense as little as a single photon. This combination of fast scintillating plastic and a sensitive light detector means we have a precise measurement of the time the particle passed through the time-of-flight scintillators—to within 100 to 200 picoseconds. That is roughly equal to the amount of time it takes light to travel an inch or two!

The last layer is the calorimeter. The name "calorimeter" dates back to devices that measured the heat or energy contents of a material, their "calories." In a detector like CLAS, it is the energy of the particle that is measured. The calorimeter for CLAS is built up of many alternating layers of lead and scintillating plastic. The lead will slow down the particle and cause it to dump its energy. It can dump its energy either by producing photons, or by knocking protons, neutrons, or pions out of the lead. The charged particles will be seen in the layers of scintillating plastic, and all the particles can knock out more particles from the next layer of lead. By this mechanism a single high-energy particle that enters the front of the calorimeter can create an avalanche of particles by the time it reaches the back of the detector. Because of this, calorimeters are sometimes also called "shower counters." The calorimeter not only tells us the energy of a particle, but because of the multiple layers of sensitive scintillator, it can also tell us what the rate of energy loss is, and that rate is a unique signature of particle type—whether it is a proton or π^+, a photon or neutron, an electron or π^-.

Although the particle has now vanished, absorbed into the calorimeter or the concrete walls of the experimental hall, the information has yet to reach its final form. CLAS has about 34,000 wires and 3,000 PMTs to read out. Each one can be digitized into pulse height and timing information. If CLAS was to light up with a complex event for an instant, with particles in all directions, it could potentially produce a half a megabyte of data! Normally, events are an order of magnitude smaller, but they happen at a rate of 1,500 times each second for eight months a year, making CLAS one of the largest producers of raw data in the world.

Yet out of the dozens of kilobytes of data from each event, all we would really like to know is what kind of particle went in which direction and with what energy. That is a factor of 1,000 reduction in the volume of data—but it still has the same information. To start with, some clever circuits within the experimental hall itself reduce the accidental data: there is no need to record and analyze the pulse height of wires that didn't fire, or wires that fired by themselves due to electronic noise or static. The remaining data are shipped out of the experimental hall over a high-speed data link to a "computer farm," which can do some initial analysis and repackaging and then forwards the data to a "data silo" until it can be fully analyzed later. The data silo is a tape drive and a robotics arm that can shuffle tapes from storage racks to the tape drive and back again. Tapes may seem old-fashioned in this age of CD and DVD data, but when calculating dollars per gigabyte of data, tape is very competitive—and with CLAS producing terabytes of data each year, that is very important.

After the experiment we can perform our analysis at a more leisurely pace. The raw data is a long list of wire ID numbers and hit times, as well as PMT ID numbers, times, and pulse amplitudes. Some parts of the analysis are straightforward. For example, if light is detected at both ends of a time-of-flight scintillator at the same moment, then we know that the particle passed through the center of the scintillator bar. Also, since we know the speed of light in plastic, we know at what time it passed through. Other parts of the analysis, like the conversion from which wire was hit to momentum, is more complex. In this analysis we make our best guess as to the initial momentum of the particle. We then simulate our guess and compare the hits in our simulation with the hits from the real event in the real detector. We then repeat the simulation with a better guess until the simulated hits are nearly identical to the real hits, in which case the momentum of our last and best simulation is essentially the momentum of the real event. Clearly our analysis depends in a critical way on how well we understand our detector, but that has just been part of the historical trend over the past century.

Each layer by itself can tell us something about the particle that raced through it, but by combining information from all the layers we can do better than the sum of the parts. For instance, by itself, the wire chamber can tell us where the particle traveled to within one wire spacing (about a centimeter). However, if we also have the timing information from the time-of-flight scintillators, we can use the drift time of the ions and calculate where a particle traveled to within a tenth of a millimeter—100 times better. With particle type and energy from the calorimeter and momentum from the wire chamber, we can infer velocity. With velocity and the time from the time-of-flight scintillators, we can trace back the particle to the original time of the scattering. With this starting time, we can do two things: we can reiterate all calculations with this precise starting time and improve our precision, and we can correlate a particle detected in one side of CLAS with a particle that came from the same beam micropulse but was detected on the other side of CLAS. It was this, the measurement of correlated particles, which really drove the design and building of this detector. Finally, we have reduced that daunting half a megabyte of data to a few simple numbers: what kind of particles were they (electrons? protons? pions?), in which directions did they start, what were their momenta and energy?

Up to this stage we have dealt with a general purpose analysis. For each experiment a unique quantity which is of special interest will have been identified. In the case of the decay of the Δ particle described in the previous chapter, it was a spectrum of counts versus the sum of the proton and pion energy. In other cases it will be a combination of angles and energies. Knowing the best combination for testing models of nucleons and quarks, much as when Bjørken convinced the SLAC people in the late 1960s to plot their data as a function of the "Bjørken-x" (see chapter 2), is a signature of a healthy and dynamic interplay between experimentalist and theorist.

The road from conceiving an experiment to mounting the effort, from the hot wire at the start of the accelerator to the data at the other end, is a long one, but it is one that is understood in minute and meticulous detail. And it is these meticulous details that give us confidence in the results: probing something one quadrillionth the size of a human is no trivial task. The data, when it is plotted as a curve in a journal, or perhaps even as a single data point, have had a long, complex, and involved history. A single point or set of points can carry with it years of labor. Yet it can also carry with it the careful and joint intelligence of a team of physicists and technical support. More important in the drive to decipher the world of quarks inside neutrons and protons, an experimental point could potentially distinguish between two competing models. However, as often as not, an experiment will not completely reject one theory or model and back another; rather, it will point to a middle ground, perhaps a combination or hybrid of the two models. Or, on those rawer and exciting occasions, it will find something new beyond the scope of either model.

- 6 -

Particle Taxonomy and Quark Soup

IF YOU'VE EVER played the old parlor game "twenty questions," you'll remember that the object is to guess some secret thing that one of your opponents has in mind. You can ask no more than twenty yes-or-no questions to narrow the possibilities before you make your guess. Traditionally, though, the player with the secret begins by answering what is, supposedly, the most primary question of all: "Animal, vegetable, or mineral?"

Whenever I played twenty questions as a child, I always found myself wondering about things that seemed to fall between the cracks—or to bridge all three categories. Is the calcium in bone "animal" or "mineral"? Where does a virus fit in—especially a computer virus? How about a plastic house plant made from hydrocarbons that come, after all, from the remains of real although ancient plants? Yet however much I might try derailing the game my aunts and uncles were teaching me, they could always pigeonhole whatever I might come up with. I never asked them where quarks would fit into this trifurcated world—but I'm sure that, without hesitation, they would have called them "minerals."

Our impulse to classify anything instantly, even if we really haven't got a clue, reflects our desire to see the world as an orderly place. And perhaps cataloging things is essential to our way of understanding: relational knowledge or comprehension by association. When we encounter something new, we need to label it, even if the label is whimsical or misguided. If we don't have a tag or pointer or index for the new object or idea, and we become distracted, our brains might not be able to find our way back to it, losing it forever. As the eighteenth-century Swedish naturalist Carolus Linnaeus told us:

> The first step of science is to know one thing from another. This knowledge consists in their specific distinctions; but in order that it may be fixed and permanent distinct names must be recorded and remembered.

Without the recording and remembering of names, things get lost. It is like trying to find a single poem in a vast library without a catalog. How did we ever find things without Dewey Decimal? In fact, the naming of objects is a truly ancient activity. In Genesis (2:19–20) Adam is given the authority to name all animals. Aristotle tried to teach us how to organize our understanding of the world in his book *Categories*. It was this passion for labeling and organizing the world that lay behind the intricate circles, rings, and terraces of Dante's *Divine Comedy*. But Dante's gazetteer did more than just name and place things as Adam had done. He also imposed an ordering, a hierarchy, on his world. Circles are subdivided into rings. The lower the ring is in the Inferno, the more heinous. The higher the terrace in Purgatory or orbit in Paradise, the more exalted.

This mania for organizing the natural world found an all-time champion in Carolus Linnaeus (1707–78). Linnaeus developed the binomial system for the classification of all living things, a system that is essentially in use today. "Binomial" implies a two-part name for every type of flora and fauna, a genus and a species, a general and a specific name, *Homo sapiens*, *Felis catus*, and *Sterna paradisaes*—the human, the house cat, and the arctic tern. Yet this is just the surface of biological taxonomy.

In the full binomial system the world is divided into "kingdoms," such as vegetable and animal. These are subdivided into "phyla," based on the most significant physical traits. Phyla are further subdivided into "class," "order," "family," "genus," and "species." For example, consider the classification of our most familiar species—the human being:

kingdom:	Animalia
phylum:	Chordata
(subphylum:	Vertebrata)
class:	Mammalia
order:	Primates
family:	Hominidae
genus:	Homo
species:	Homo sapiens

But if we viewed the binomial system as only pigeonholes for species, "a place for everything, and everything in its place," that would trivialize the role of a good cataloging system.

In this system, catalogers understood that important traits come in the higher levels of the hierarchy. For instance, Linnaeus recognized that "chordata" is an important trait that distinguishes a large group of animals—those with a notochord, or central nerve, running along the back. Hence, chordata is a phylum and very near the top of the system. In contrast, the difference between an arctic tern (*Sterna paradisaes*) and a common tern (*Sterna hirundo*)—as Roger Tory Peterson in his field guide tells us—is a small band of white near the bird's beak. This small distinction, or "field mark," is the only visible trait that stands between the two species. Therefore, the two terns are classified as part of the same genus.

A second feature of a good classification system is that it gives us a glimpse of the underlying axioms of the system. In chapter 2 we saw that the periodic table of the elements of Dmitri Mendeleev had its structure—unbeknownst to Mendeleev—because of the way electrons can occupy atomic orbits. The first row of the periodic table has two elements because the first orbit of an atom can have two electrons. The elements of the first column, the alkali metals, have one electron in their outermost orbit. The noble elements, in the last column, have their outermost orbits completely filled.

In a similar manner, Linnaeus's binomial system has its structure dictated by that all-encompassing theory of biology: evolution. The tree structure of the full taxonomic system is a shadow of the tree of evolution. By this I mean that the evolutionary path of two species of different phyla diverged eons ago, whereas the separation between the arctic tern and the common tern is an event of yesterday. Generally speaking, similar traits arise from similar genetics and speak of a similar history.

So now, when we turn from the vernal world of biology to the quiet and deep domain of the nucleons and quarks, we might be tempted to lay out a "binomial system"—or a "periodic table"—of particles. But learning a list of particular traits instead of starting with the underlying elements or axioms is a long road. If it is done with any detail, it is a recounting of the history of particle physics, complete with false starts and blind alleys. It would be like learning every physical trait of a myriad of species before learning about Darwin's theory of evolution. Or like learning all the properties of all chemical elements—what every alkali and transition metal look like—before discussing the electron and the proton. Following this path would take us along the same route as Gell-Mann, whom we discussed in chapter 2. And still, in the end we would end up with the patterns that Gell-Mann observed and named the "eight-fold way."

Instead, let us start with the underlying theory of quarks and show how we can derive the particular traits of the observed particles. Starting with half a dozen different types of quarks and a few rules of combinatorics, we can cook up the hundreds of observed particles. In the end, we must be able to reproduce the "eight-fold way," as well as the vastly more complex and extended patterns that arise from new particles discovered in the three and a half decades since Gell-Mann's paper.

There are six ingredients to quark soup: the "down" and "up" quarks, the "strange" and "charm" quarks, and the "bottom" and "top" quarks. They all have spin $\frac{1}{2}$, an electrical charge of $+\frac{2}{3}$ or $-\frac{1}{3}$, and a new and curious quantity called color charge. Each of the six types, or "flavors" as they are generally called, has a distinct mass and a preferred decay pattern. And that is all the properties of quarks—they are very simple objects.

The labeling of the types of quarks as flavors may seem particularly whimsical, especially when compared to the classical names used by previous generations of physicists, such as "atom," "electron," "proton," and "neutron." Even the particles such as the τ, the Δ, the ρ, and the Λ, have names that tie themselves to the world of Democritus. But the nomenclature of quarks is quite different. Maybe it was an artifact of the 1960s and 1970s, or perhaps it was merely meant to emphasize the fact that there is nothing in our common and macroscopic experience to guide us. However, the names are not completely arbitrary. They are designed to help us remember which quarks tend to be paired together. We call the pairing a "family." The up and down quarks, or *u* and *d* quarks in the physicist's shorthand, are the first and most familiar family, and account for the most common particles, the proton, neutron, and pion. As we progress along the list, the quarks are more massive. Therefore, they require more energy to be formed, they have shorter lifetimes, and they become more and more rare and elusive. The second family is made up of the charm and strange quarks. Finally, the third family, the most massive and the most recently "discovered" quarks, are the top and bottom quarks.

The names "up quark" and "down quark" follow from the older nuclear physics models in which the proton and the neutron are treated as if they are the same particle (the nucleon) with a new quality called isospin distinguishing the two. Thus, the proton is described as a nucleon with "isospin-up" and the neutron is a nucleon with "isospin-down." The advantage to this model was that nuclear physics could then use all the mathematics that atomic physics had developed for electrons with spin. Therefore, when the quark model was being developed it was realized that the proton had to have more "up-ness" and the neutron more "down-ness," so these quark names naturally followed.

The case of the strange quark also evolved out of the preexisting conventions. The kaon (K) and the lambda (Λ) had been observed in the 1950s and called strange because of their unexplained appearance and semistability. In fact, many of the particles made up of the three lightest quarks were all well known in 1964, and Gell-Mann even used the notations u, d, and s in his first paper on quarks. However, the idea of families was not at all developed, and the name "strange" was not designed as part of the doublet. In 1964, Bjørken and Glashow came up with the name "charm," to complement strangeness.

"Top" and "bottom" were originally proposed as "truth" and "beauty," but these names seem to have fallen into general disuse. Perhaps "truth" and "beauty" were just too whimsical, or perhaps they really did not form a pair of opposites:

> Beauty is truth, truth beauty—that is all
> ye know on earth, and all ye need to know.
>
> *Keats: Ode on a Grecian Urn*

There was even a movement at one time to replace the names with only the letters t and b, but "top" and "bottom" seem to be the stable middle ground. Yet names are merely labels and not the "specific distinctions" Linnaeus would have us seek. For these distinctions we turn to the physical traits: the masses and the decay patterns.

The masses of the different flavors are distinct, but extremely difficult to measure. The usual method for measuring the mass of a particle involves isolating it and measuring its momentum, energy, or velocity. By knowing two of these three quantities, mass can be derived. This method works well for measuring the mass of a pion or proton or kaon, but it doesn't work so well for a quark, for we cannot isolate a single quark. This is the unique problem in studying the hidden world of quarks inside protons and neutrons. We must depend on theories.

We start out with a quark theory and then calculate experimental observables, such as the cross sections or masses of macroscopic particles like the proton or the pion. We then adjust the masses of the quarks and various other parameters in the theory until our calculations agree with the laboratory measurements. This sounds like a technique that guarantees success, as a cookbook might tell us, "add seasonings to taste." Add and adjust parameters until the theory describes nature. But this is where "Occam's razor" comes into play: the simplest theory that describes the widest range of observations is the preferred theory. Also, there is a sense that the parameters of the theory must be associated with something in nature itself. A theory whose parameters are masses and charges would be preferred to a theory of nonphysical parameters (a, b, . . . , z), although, admittedly, the whole quark theory was initially susceptible to this criticism.

Still, all of these criteria do not uniquely identify a single quark theory. In fact, there are two different theories in general use, and the masses of the quarks derived depend on the theory being used. The two theories are QCD and the constituent quark model. QCD is almost certainly the best theory we have of quark dynamics in terms of the ultramicroscopic structure of the subnucleon world, but it is a complex and difficult theory with which to work. The quark masses used in QCD are called bare masses, or current masses. In contrast, the constituent quark model is much simpler, using equations similar to Schrödinger's equation for the atom. However, the constituent quark model has a much lower status than QCD, as indicated by its name. QCD is a theory with a truly fundamental foundation, whereas the constituent quark model is merely a model—useful for calculations, but not a fundamental theory of matter.

It is not that we are so ignorant about quarks as to entertain two opposing theories. Rather, they are complementary theories with their own appropriate applications. It is like comparing the atom seen by a chemist with that seen by a physicist. The chemist might see the atom as a solid object with the ability to bind to its neighboring atoms to form molecules. A physicist might see the atom as electrons and a nucleus with a lot of space in between, and would also describe the atoms in terms of orbits or energy levels. Admittedly, the orbits have a great deal to do with what chemical bonds are formed, but the two views of the atom are both useful and have their distinct and preferred scales and scopes. The same is true of QCD and the constituent quark model.

All qualifiers aside, the constituent quark model is still very useful. In this model the effects of the gluons (we will talk about gluons in detail later) that swarm around each QCD bare quark are rolled up with that bare quark to form a larger quark—a "constituent quark," with a "constituent-quark mass." We can estimate the constituent masses in a simple way. First, the proton is built up of two up quarks and one down quark ($p = uud$), whereas the neutron is made of one up quark and two down quarks ($n = udd$). So the difference in masses of the two nucleons must be related to the differences of the masses of the up and down quarks. In fact, since the neutron and the proton have almost the same mass, the constituent masses of the up and down quarks must be almost identical. We can then also conclude that each quark must have roughly a third of the mass of the nucleons. The masses from QCD are much lighter, since in QCD a great deal of energy is attributed to the gluons, to the binding between quarks, and especially the binding of quarks to themselves.

There are more quarks than just the up and down quarks, and fortunately nature has provided us with a great many more particles and masses than just the neutron and the proton to guide us. With dozens of particles observed, we can estimate the mass of the quarks as:

	constituent-mass (GeV/c^2)	bare-mass (QCD) (GeV/c^2)
up	~ 0.3	0.001–0.005
down	~ 0.3	0.003–0.009
charm	~ 0.7	1.15–1.35
strange	~ 0.5	0.075–0.170
top	—	169–179
bottom	~ 1.2	4.0–4.4

The fact that we can start with only six quarks and build essentially all the observed particles is a compelling argument in support of a quark theory—but it is only part of the macroscopic evidence, the evidence we see from the outside particles. In addition to mass, each quark has a unique charge and a distinct decay pattern.

The peculiar fractional charges of quarks were one of the original unpalatable features of the quark model. The up, charm, and top quarks have a charge of $+\frac{2}{3}e$ ($-1e$ is the charge of an electron) and the down, strange, and bottom quarks have a charge of $-\frac{1}{3}e$. In the early 1960s, this contradicted common sense. Ever since Millikan's water and oil-drop experiments we had viewed the world as being made of integer charges, $+1e$, 0, and $-1e$, and nothing else. But we had never, and still have never, directly seen a fractional charge. We now understand this in terms of quarks *always* coming in groups of three or in quark-antiquark pairs. Given that they must be bound into these combinations, perhaps it is not so strange that these fractional charges remain concealed in the stable, observed particles, but what is less clear is why fractional charges should remain elusive after particles fall apart and decay.

To understand the inner mechanics of a decay let us turn to the quintessential weak decay, the beta decay of a neutron into a proton, electron, and neutrino. On the quark level the decay looks like the diagram shown in figure 6.1. The decay of the macroscopic particle, the neutron, is a result of the decay of one quark, the down quark. The other two quarks are referred to as "spectators." (In the much older terminology for describing the ordinary chemistry of substances dissolved in water, ions that do not participate in a reaction are called "spectator" ions.) This simple neutron beta decay is perhaps the most common of decays, yet buried in it is a wealth of quark physics, if only we can unravel it. First, charge is conserved. That is one of the great canons of physics: charge is always conserved. Also, when a quark decays, it always decays into one quark and a short-lived particle called the *W*-boson, which then immediately decays into some-

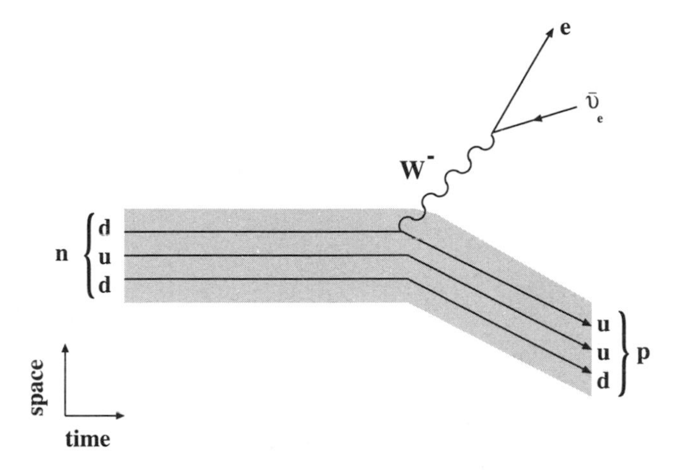

Figure 6.1 Beta decay—the quintessential weak decay.

thing else (in this case, an electron and neutrino combination). This means that when a quark decays into another flavor, the magnitude of its charge can change by only the charge of a *W*-boson, that is, by an amount equal to or opposite that of the charge of an electron (there are both positive and negative versions of the *W*-boson). As a result of this chain of reasoning, we find that it is not the charge of the quarks that must be proportional to the electron charge; rather, it is the *difference* between the quark charges that is $\pm 1e$.

One of the curious properties of a beta decay is that it is modified by its environment. The rate of decay of the down quark into an up quark is slowed because the quarks are embedded inside a nucleon. We would like to look at an even simpler case, the decay of a free quark. But of course there is no such thing. Nature, however, does provide us with a telling phenomenon: the beta decay of a muon (figure 6.2). The muon is just like an electron, except it is 200 times heavier. Its beta decay is very clean, and the theoretical predictions and experimental results are in exceedingly good agreement. Experimentally and theoretically, the lifetime of the muon is 2×10^{-6} second, or 2 milliseconds. That may seem fast, but it is forever compared to the strong decays, which limit the lifetime of a Δ particle to 5×10^{-21} second—the

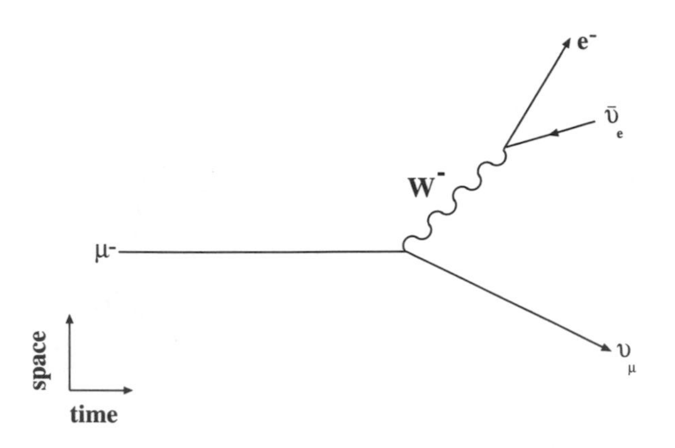

Figure 6.2 Muon decay—the simplest of all weak decays.

time it takes light to cross two protons. To try and understand these scales, if we expand the lifetime of the Δ particle to a single second, the muon would prosper for 13 million years!

We return now to the neutron, with which we started our discussion of decays. It has a lifetime of 14 minutes. When we scale the Δ's lifetime to one second, giving the muon a ripe old age of 13 million years, the neutron becomes more ancient than the stars, with a lifetime of 5 quadrillion years—about half a million times the age of the universe! This longevity arises primarily from the fact that the masses of the down and up quarks are so similar, which means that there is little energy available to drive the decay.

A second factor that determines lifetime is that quarks do not decay into all other flavors with equal probability, even after we have adjusted decay rates for different masses and allowed energies. Instead, a quark will tend to decay into the flavor of its sibling, the flavor of the other quark in the same family. Quarks can also decay into flavors of the adjacent family, and on rare occasions there may even be decays between the first and third families. Schematically we can lay out the allowed decays as in figure 6.3. The decay between family members is the preferred direction; between adjacent families is possible; and between distant families is a rarity.

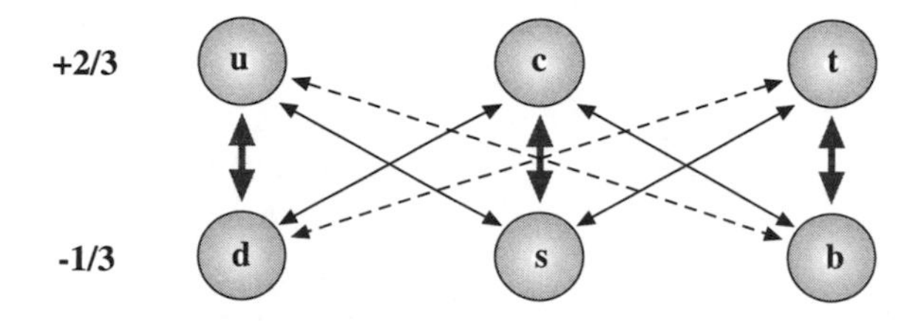

Figure 6.3 The allowed weak decays. The thick arrows indicate the most likely decays: (*u-d*), (*c-s*), and (*t-b*). The thin arrows indicate decays that take place occasionally: (*u-s*), (*d-c*), (*c-b*), and (*s-t*). The dotted arrow indicates the rarest of decays (*u-b*) and (*d-t*).

One last subtle detail is that when the down quark in the neutron decays into an up quark in a proton, the quark must be in an allowed orbit in *both* situations, when it is in the neutron and when it is in the proton. But where those orbits are and how the quarks arrange themselves within nucleons we will leave for the following chapter.

Since we now know the quark charge, mass, allowed decays, and decay rates, let us try to build up a particle and see how well it compares to nature. For example, if we combine the three lightest quarks, the up, down, and strange quarks, their masses add up to 1.1 GeV/c^2. The normal decay mode will be for the strange quark to decay into an up quark and a *W*-boson. The boson will then decay into a anti-up quark ($\bar{u}$) and a down quark (d). Finally, the four quarks and one antiquark that come out of the decay can pair themselves up in two different ways, giving rise to two distinct sets of daughter particles, a (π^0 n) or (π^- p) pair. The Feynman diagram for these two alternative decays is shown in figure 6.4.

In fact in nature we find a particle with a mass of 1.115 GeV/c^2, which decays into (π^- p) or (π^0 n). It is called the lambda (Λ). But to understand the decay of the Λ in detail we must introduce the last property that quarks possess: "color." If electrical charge ($+e$/$-e$) is the charge that binds electrons and the nucleus to form atoms, then the three color charges ($r/\bar{r}$, $b/\bar{b}$, $g/\bar{g}$) are what bind quarks together to form a nucleon, or any hadron. As we saw in

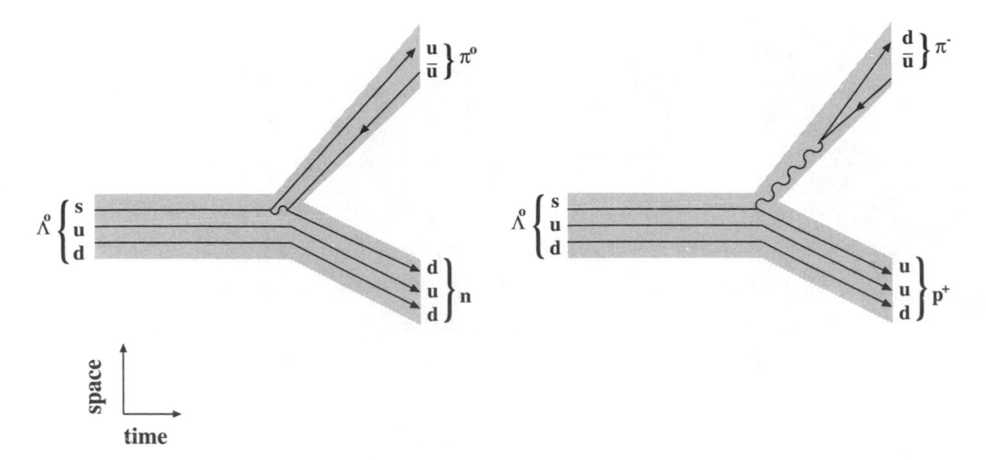

Figure 6.4 Lambda decay. The Λ can decay into a (π^0 n) or a (π^- p) pair. The second combination is more likely, since the quarks in the π^- can have any color.

chapter 3, the rule is that a baryon (three-quark combination) will be made up of the three colors (*rbg*), whereas mesons (quark-antiquark pairs) will be made of color-anticolor pairs ($r\bar{r}$ or $b\bar{b}$ or $g\bar{g}$). The effect on the decay diagrammed above is curious. Let us initially randomly assign colors to the original three quarks. Say that the strange quark is red, the up quark is blue, the down quark is green. In the decay into the π^0 and n, all the colors of the final quark are now determined. Since a quark keeps its color when it decays, the up quark in the π^0 is red, its partner must be anti-red, and the new down quark is also red. But in the second diagram, the colors of the $\bar{u}$ and d in the π^- are completely arbitrary. We should replace the second diagram with three different colored diagrams, for the π^- could be made up of $r\bar{r}$, $b\bar{b}$, or $g\bar{g}$. Now the probability of a lambda decaying into a $p\,\pi^-$ is not just 3 times larger than the probability of decaying into a $n\,\pi^0$. Rather, it is enhanced by only the square root of 3, for the same reason that probabilities are the square of wavefunctions in quantum mechanics. Finally, from our color counting we have calculated that the "branching ratio" of lambda is given by:

$$\frac{\Lambda \rightarrow p\,\pi^-}{\Lambda \rightarrow n\,\pi^0} = \sqrt{3} \approx \frac{65\%}{34\%}$$

Lambdas decay into a $p\,\pi^-$ 65 percent of the time and a $n\,\pi^0$ 36 percent of the time, in excellent agreement with experimental results.

The point of this digression into the decay of the lambda is that we can take any combination of quarks and then predict the particle's mass, spin, charge, lifetime, and branching ratios of the combinations. We can also apply this sort of quark addition to the mesons with great success. A quark and an antiquark can combine into a meson. With six flavors of quarks, and six antiquarks, we would expect thirty-six basic mesons, but the combinations involving the top quark are so short-lived that they have not been observed or named. The other twenty-five are listed below:

	$\bar{d}$ (+1/3)	$\bar{u}$ (−2/3)	$\bar{s}$ (+1/3)	$\bar{c}$ (−2/3)	$\bar{b}$ (+1/3)	$\bar{t}$ (−2/3)
d (−1/3)	π^0	π^-	$\bar{K}^0$	D^-	$\bar{B}^0$	
u (+2/3)	π^+	π^0	K^+	$\bar{D}^0$	B^+	
s (−1/3)	K^0	K	ϕ	D_s^-	$\bar{B}_s^0$	
c (+2/3)	D^+	D^0	D_s^+	J/ψ	B_c^+	
b (−1/3)	B^0	B^-	B_s^0	B_s^-	Υ	
t (+2/3)						

An anomaly that immediately catches the eye is that both the $(u\bar{u})$ and the $(d\bar{d})$ combine to make up a π^0. In reality, it is more the other way around, that is, the π^0 is made up of a combination of $(u\bar{u})$ and $(d\bar{d})$ pairs. This is because ups and downs have nearly identical constituent masses and therefore they can "decay" into each other—a $(u\bar{u})$ pair can become a $(d\bar{d})$ pair and likewise a $(d\bar{d})$ pair can become a $(u\bar{u})$ pair. We can imagine that inside a π^0, a series of decays is always taking place, as shown in figure 6.5. Also, even if the $u\bar{u}$ were not "entangled" with the $d\bar{d}$, there is no experimental way of distinguishing the two. There is no unique observable, no field marks to delineate the two.

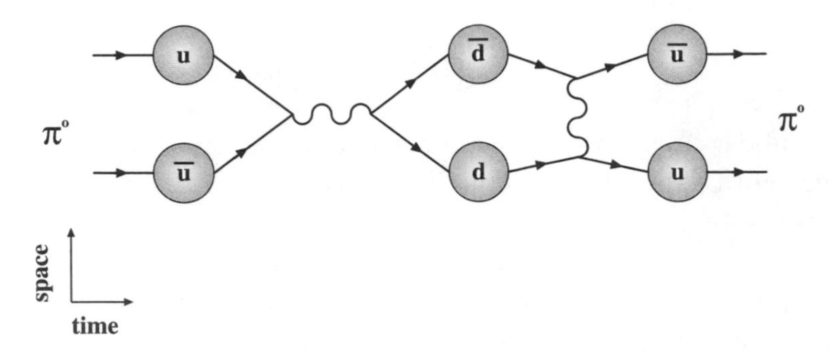

Figure 6.5 The life of a π°. The π° can oscillate between being a combination of a $u\bar{u}$ or a $d\bar{d}$ pair of quarks.

So far we have discussed such field marks as charge, mass, and weak decays, but there is one more observable permutation of quarks that we encountered in earlier chapters: spin. When the spin of a quark in a nucleon flips, the particle gains mass and we call it a Δ particle. The possibility of spin flips are universal in the quark world, and the process can be applied to any of the quark-antiquark pairs listed above. In the mesons we have listed, the spins of the quark and antiquark are antiparallel, that is, they point in the opposite direction. This means that when we combine the spins of the quarks to get the overall spin of the meson, the spins cancel. So the spin of the meson is zero, and therefore the particle has no preferred direction. These spin-zero mesons are collectively referred to as pseudoscalar mesons. Still, there is the possibility of aligning the spins of the two quarks to add up to spin 1. This creates a more massive meson, called a vector meson. Each pseudoscalar meson we have listed has its heavier alter-ego vector meson:

$$\begin{array}{lcl} \pi(140) & \leftrightarrow & \rho(770) \\ K(500) & \leftrightarrow & K^*(890) \\ D(1860) & \leftrightarrow & D^*(2000) \\ B(5280) & \leftrightarrow & B^*(5325) \end{array}$$

The name "pseudoscalar" describes the observable consequence of having spin 0. With no spin, these mesons have no preferred direction. If the meson scatters off something, it scatters equally in all directions. If it decays, its daughter particles fly off in all directions uniformly. Vector mesons have an orientation, a preferred direction given by their spin orientation. So when a vector meson scatters or decays, the angular distribution of the scattered particles, or daughter particles, is a unique signature, a field mark, of its "vectorness."

It seems that every time we think we have a handle on the number of ways to add up quarks, and thus the number of combinations, or macroscopic particles, I yell out, "But wait, there is more." First it was more flavors—each with its special mass, electric charge, and decay pattern. Then we added in color charge, which changed our counting rules. Next we added in spin, which doubled the number of particles. But still, we have not described the most basic and familiar particles, the neutron and the proton. So let us finally turn to the last permutation of the quark adding rules, the addition of three quarks to make a "baryon." At first glance we might expect that there are $6 \times 6 \times 6 = 216$ different baryons, since each quark can have one of six flavors. But the mathematics is not only about combinations, it is also about permutations. For example, a proton is not just a (*uud*), but also (*udu*) and (*duu*). Then the original list of 216 combinations reduces to fifty-six real combinations of quarks. With what we learned from mesons, and paying careful attention to the Pauli exclusion principle, we can essentially derive all the stable baryons that physicists have observed.

Still, to work through fifty-six combinations of quarks is a great deal more drudgery than necessary, and we can learn everything we need to know about three-quark combinations by considering only the three lightest and most common quarks: the up, the down, and the strange quarks. First, there are ten unique combinations of these three quarks (as listed below). Also, as we pointed out when trying to understand the Δ particle, there are two unique spin combinations: spin $\frac{1}{2}$, when one quark is antiparallel with respect to the two other quarks, or spin $\frac{3}{2}$, where the spins of all the quarks are aligned. We can then tabulate the bary ons made up of the three lightest quarks and spin $\frac{1}{2}$ or $\frac{3}{2}$:

combinations	spin 1/2	spin 3/2
ddd		Δ^-
ddu	n	Δ^0
uud	p	Δ^+
uuu		Δ^{++}
dds	Σ^-	Σ^{-*}
uds	Σ^0 Λ	Σ^{0*}
uus	Σ^+	Σ^{+*}
dss	Ξ^-	Ξ^{-*}
uss	Ξ^+	Ξ^{+*}
sss		Ω

There are three things about this table that grab our attention. First, the (*uuu*), (*ddd*), and (*sss*) spin $\frac{1}{2}$ combinations are missing. Second, there are two particles for the spin $\frac{1}{2}$ (*uds*) combination. Third, these are the baryons that showed up the Gell-Mann's "eight-fold way" paper, which I mentioned in chapter 2.

The (*uuu*), (*ddd*), and (*sss*) spin $\frac{1}{2}$ combinations are missing because if these particles existed they would be fermions, but would not satisfy all the criteria required for fermions. Fermions are particles that have half-integer spins ($\frac{1}{2}, \frac{3}{2}, \frac{5}{2}, \ldots$) and quantum mechanics requires that fermions must have "antisymmetric total wavefunctions." In a sense, this is the generalization of the Pauli exclusion principle. But why are these combinations "antisymmetric"? The reason goes back to color. When we introduced an antisymmetric "color wavefunction" on top of flavor and spin, the combination (*uuu*) (which would be a p^{++}—a doubly charged proton) would be antisymmetric with respect to both spin and color, which means symmetric with respect to the product of flavor, spin, and color, and so therefore not allowed. Although it was the observation of the Δ^{++} and the absence of the p^{++} that introduced the quantity color, its presence is far-reaching, from such a phenomenon as the decay of the Λ to the fact that baryons are always made up of three quarks. But what color is and how it combines will have to wait for the following chapter.

The final phenomenon we must explain is the fact that the three-quark combination (*uds*) can form either a Λ or a Σ^o particle. To understand this, let us build it up, piece by piece. We can build it out of $u\downarrow d\uparrow s\uparrow$ or $u\uparrow d\downarrow s\uparrow$ or $u\uparrow d\uparrow s\downarrow$. The up and down quarks are so similar that to understand the symmetry of the whole particle, we can start out by just looking at the symmetry of these two. In the first two combinations we can combine in an antisymmetric way to get the Λ particle :

$$\Lambda = [\ (u\downarrow d\uparrow) -- (u\uparrow d\downarrow)\ ]\ s\uparrow$$

This is "antisymmetric" because if we exchange the up quark and the down quark we change the total sign of the combination ($[u\downarrow d\uparrow - u\uparrow d\downarrow]\ s\uparrow = --[d\downarrow u\uparrow -- d\uparrow u\downarrow]\ s\uparrow$). In fact, getting these two combinations of ups and downs into the same baryon is the same mathematics as getting two electrons into the same orbit of an atom. The remaining combination, $u\uparrow d\uparrow s\downarrow$, will have a total isospin of 1 and is identified as the Σ^o. The Λ and the Σ^o each have their own unique mass. The Σ^o is about 10 percent higher than the Λ, but the real field mark is the decay. The Σ^o has that unit of isospin it must get rid of when it decays. So it decays by giving up an energetic photon and changing into a Λ!

It is easy to lose the overall picture of the quark model of mesons and baryons when we are caught up in a long list of observable phenomena. We want to think about the quark model as an ecliptic description of all hadronic physics, but we test it one observable at a time. It is the details of the observables, all those hundreds of particle masses, decay rates, and cross sections, which build confidence in the model. But it is the fact that we can start with such a few axioms—six flavors, two spins, and three colors—and produce so much, which gives the theory its great appeal. Although we cannot see the quark, we can see—and measure—hundreds of observables that the theory has predicted.

Perhaps a decade ago the theory lost the flavors truth and beauty in favor of the less colorful top and bottom. But the theory retained its beauty where it was more important, in its expansive scope and its simplicity. Science may only speak in terms of "confidence," and absolute knowledge may be denied us, but the mounting evidence strongly suggests that the quark model of subnuclear particles is very close to the truth.

- 7 -

The Shape of Things

IF YOU TAKE a close-up look at a mosaic and then step away and watch as thousands of individual tiles merge into a single image, the effect is nothing short of amazing. One might even argue that the true beauty of a Byzantine masterpiece is not so much in the larger image as it is in the magic of that transition from tiny, dull, repetitive pieces into a large and coherent picture. Clearly the whole is something quite different from an unsorted collection of the parts. The mosaic has properties that the tiles alone do not possess. Rather, it is the macroscopic arrangement of the tiles that gives life to the image.

This distinction between the properties of a system that are derived from its parts, and new properties that arise from the arrangement and interplay of the parts, is manifest at many levels in the world. To take a second example, consider a system with only a few members: a musical trio. If one member is ill, and only two musicians step out onto the stage to play two-thirds of the music, there is clearly something missing. The music was composed for three and the absence of one performer will overshadow the presence of the other two, because a composition designed for three musicians doesn't reduce to a duet when one part is missing. A trio is something more than just three musicians who are sharing a common stage. An unsynchronized performance would lead to an awful din, like the sounds we hear during a warmup before a performance. A concert is something very different. A concert takes place when the musicians consort, when they coordinate and wrap their airs around each other's, and produce something, a whole that is much more than the sum of the parts. Even in a system as simple as trios and duets, the coordination of the individuals can lead to properties not inherited from the individuals alone.

One final example, before we return to the quarks and nucleons, can be found in the study of the atom. At the beginning of the twentieth century, J. J. Thompson—the discoverer of the electron and the director of Cavendish Laboratories—described the atom as a kind of amorphous pudding, with the occasional plum or raisin embedded in it. This structureless "plum-pudding" atom had some of the macroscopic properties that J. J. and his contemporaries were aware of: it had the right charge, and the plums were but a small part of the whole. Yet the plum-pudding model missed the details of the atom. It was like hearing a trio and missing the fact that the musicians are playing in concert. It was left to Bohr and his atomic orbits, as well as Schrödinger and his wavefunctions, to tell us that the atom has a unique structure. It is this structure that gives rise to not only the atomic spectrum, but also the unique chemical properties of each element, the macroscopic traits by which we know oxygen from carbon even at a glance.

So what is the structure of protons and neutrons? How do quarks arrange themselves inside a nucleon? At the time of this writing, the answer is not crystal clear. In the past two decades there have been some experiments that have made initial measurements. But perhaps the most important results of these experiments are that they have prodded theorists to study the arrangement of quarks in nucleons, and they have identified the experimental techniques we will need to develop. They have told us that to probe the structure of a nucleon we will need a combination of accelerators that can produce polarized beams and targets made of polarized nucleons.

However, before cutting steel to build a new detector, or pouring concrete to build a new laboratory, we can learn a lot about the structure of a nucleon by finding the most quiet and contemplative corner of the library, or by taking the most desolate and secluded walk and asking ourselves, How might they arrange themselves? This amble may seem unproductive, but remember that the Greeks "proved" the existence of the atom, and the boundless extent of the universe, with no more apparatus than their brains and a quiet olive grove.

To begin with, set out three pebbles (or periwinkle shells, or olive pits) in front of yourself and ask, How might these three objects arrange themselves? Their first characteristic is so elementary that the tabletop where the pebbles lie may disguise it. Three objects on a table, and three quarks in a nucleon, will arrange themselves in a plane. But this is not a surprise, we have all known this since our first exposure to geometry, "three non-co-linear points define a plane." The other thing three points define is a triangle, but beyond these general statements there is a lot of freedom. A triangle might be equilateral, scalene, or obtuse. It might have two points a fermi apart, and the third point in the next galaxy—but that triangle has little to do with a nucleon. So let us add some physics to the mix.

The binding of the quarks depends primarily on the color of the quarks, and then on their spin, but the binding doesn't depend on the flavor—or type—of the quarks. QCD tells us that a red up quark and a blue down quark are bound as tightly as a red up quark and a blue up quark. This "flavor-independent" force is powerful, and very curious. If the two quarks are near each other, they experience almost no attraction. This is referred to as "asymptotic freedom." However, as they move farther and farther apart, the binding force rises to 22 tons! In other words, if you could grab a single quark and leave its partners in the top of a full dump truck, you could lift the truck up into the air before the quark-quark bond broke. However, we are less interested in this extreme case, and more interested in the situation where the quarks are near each other, as in a nucleon. Since a nucleon is made up of three quarks, each with essentially the same mass and all with the same binding, they will form a most democratic configuration; they will arrange themselves into an simple equilateral triangle—if that were the whole story.

The QCD force between quarks does not depend on flavor, but it does depend on spin, and it is spin that breaks the symmetry of that democratic triangle. Perhaps this is not so surprising, for we have already seen the dramatic effect of spin when comparing the nucleons with the Δ particles. If a spin flip can change the mass of the three-quark system by 30 percent, the fact that it has a role in the shape of the nucleon seems most natural. Curiously enough, we can derive the characteristic shape of the neutron on the basis of two basic principles, and a bit of logic and reasoning that most quiet olive groves could precipitate. The two principles are old and familiar ones: (1) opposites attract and (2) the Pauli exclusion principle.

Let us start with what appears to be the most mundane and bland of particles, the chargeless neutron. What do we really know about it? Macroscopically it has a spin of one-half; it has an electric charge of zero; and it is roughly the size and mass of a proton, a fermi across, and 940 MeV/c^2 heavy. Microscopically it is built up out of three quarks: two down quarks and one up quark. In order for the neutron to have a total spin of one, the spins of two of these quarks must be aligned and pointing in the same direction, whereas the third quark is oriented with its spin in the opposite direction. And that really is all we know to start with.

But which quark has which spin? This is where the Pauli exclusion principle comes in—but with an unexpected twist. The Pauli exclusion principle tells us that two identical particles must be "totally antisymmetric." It doesn't tell us anything about any particular type of symmetry; rather, it only tells us about the product of all the symmetries that the system has. In the familiar atomic and nuclear physics cases this means that if two particles are identical or symmetric in all other properties, then that leaves only their spins to be "antisymmetric" or opposite. Experience reminds us that two electrons in the ground state of an atom will have opposite spins in compliance with the Pauli exclusion principle. In fact, it seems as if the helium-3 nucleus can provide us with a perfectly analogous case to compare to three quarks in a neutron. The helium-3 nucleus is made up of two protons and one neutron. The protons and the neutron have spin $\frac{1}{2}$, as does the whole nucleus. We therefore know that two of the nucleons must have parallel spins, and one of them has its spin opposite. The analogy with three quarks seems nearly complete. In helium-3, the two protons are identical in every respect, with only the spin orientation left to be determined. Therefore, the Pauli exclusion principle *requires* that they have opposite spins. So the spins of the protons effectively cancel each other out, and the spin of the whole nucleus is essentially determined by the spin orientation of the neutron! Although this is a useful feature, as we will see later, it is a misleading and deceptive analogy when we turn to the world of quarks inside a neutron.

Now, let us look at the world with a more powerful microscope, and consider the arrangement of three quarks. As I have listed before: in a neutron we have an up quark and two down quarks. The down quarks have identical charge, mass, and flavor, but they are antisymmetric in color. Color is a most curious quantity. It was originally proposed shortly after the advent of the quark model to explain how three up quarks could combine into the Δ^{++} particle. We have, up to now, described the Δ particle as a proton or a neutron with spin $\frac{3}{2}$. We understood this in terms of one quark having its spin flipped. All this it true, but what I neglected to develop was the two other Δ particles. All told, there are four Δ particles. There is the Δ^{++}, made up of three up quarks (*uuu*). The Δ^{+} is the counterpart to the proton and, like the proton, is made up of two up quarks and a down quark (*uud*). The counterpart to the neutron is the Δ^{0} (*udd*). Finally, there is the Δ^{-} (*ddd*). These appeared in the baryon-decuplet diagram in chapter 2 (figure 2.3) and the baryon table of chapter 6. But one of the original problems for the quark model to explain was the symmetry of the Δ^{++} and Δ^0. It was clear that the three quarks in the Δ^{++} were three up quarks, each with their spins oriented in the same direction. They were symmetric in every respect that physicists were aware of: spin, mass, flavor, and even location. This symmetry crisis is what led to the original postulate of another quantum quantity, or "quantum number," called color. Color could be arranged in such a manner that the Δ^{++} was antisymmetric in color, and then the Pauli exclusion principle would be satisfied. Now at first this seems like a contrived solution. To postulate colors that you can't see to solve the problem of symmetry seems on a par with the medieval postulate of celestial spheres that were too clear to be seen, or the Victorian theory of ether that was proposed to mediate light, yet itself was too ethereal to be detected. But color is different. Its effects are seen and measured in other ways. For instance, color is the reason a Λ particle tends to decay more often into a proton and a negative pion ($\Lambda \rightarrow p\,\pi^-$) instead to the neutral pair of neutron and neutral pion ($\Lambda \rightarrow n\,\pi^0$), as we saw in the previous chapter. In fact, color redistributes dozens and dozens of decay processes, and color is now understood far beyond its role in symmetries. But color's role in symmetries is still quintessential. It means that if the first down quark in a neutron is blue, the second down most certainly cannot be blue. It can be red or green, or even a quantum mechanical combination of red and green—anything but blue.

So back to the problem of the neutron. If the two down quarks in the neutron are already color antisymmetric, where does that leave their spin? The spin must be *symmetric* to satisfy the Pauli exclusion principle, for the Pauli principle is about the *total symmetry* of the whole system. If the down quarks were antisymmetric in color *and* spin, the sum total for the whole neutron system would be symmetric—which is a forbidden combination. This seems like a long argument just to establish the fact that the spins of the two down quarks in a neutron are symmetric or parallel, but this fact will drive the shape of the neutron.

The last piece that we must introduce is that "opposites attract," and "the same repels," something we learned with mag-

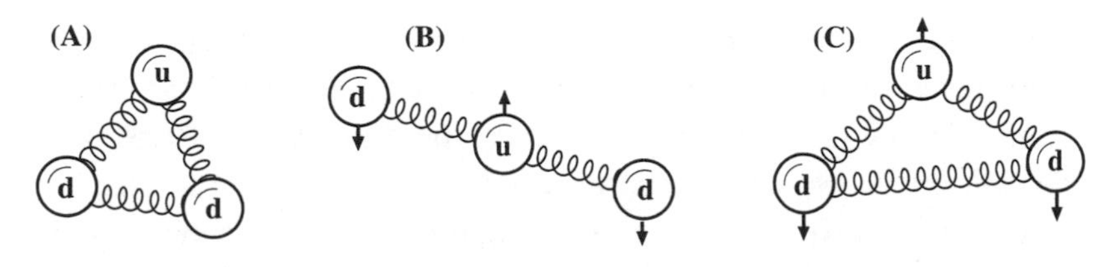

Figure 7.1 (A) If the force between quarks were independent of spin, quarks would arrange themselves in an equilateral triangle. (B) If the force between quarks depended only on spin, the quarks would be arranged in a line. (C) The real force between quarks has both a spin-dependent and a spin-independent component.

nets in elementary school. It seems absurd that a principle that is true for bar magnets on a tabletop, and that orients a compass thousands of miles from the Earth's poles, could also apply to quarks spinning in a nucleon. But the mathematics of combining forces that arise from a pole, be it a magnetic pole or a spin axis, is quite limited, and the equations that describe magnetic interaction look essentially the same as the ones that affect quarks. So the two down quarks, with parallel spins, have a degree of repulsion. However, it is true that among quarks the spin-dependent interaction is not the major force. In fact, it plays a secondary, but significant, role, after the color force.

To recast the problem, if the binding between quarks depended only on color, a force that is identical among all quarks, then the quarks in a neutron would be pulled into the most symmetric configuration: an equilateral triangle. On the other hand, if the binding of the quarks only depended on spin, the quarks would arrange themselves in a straight line with the up quark in the middle. This is because the up quark has its spin opposite to the spins of both of the down quarks, and the up-down quark spin interaction is attractive (opposites attract) whereas the down-down quark spin interaction is repulsive (same repel). In reality we expect the true neutron configuration to be a combination of these two geometries, someplace between the equilateral triangle and the three quarks in a row (see figure 7.1). The average distance between quarks tells us something about the color-dependent binding, and the distortion of the triangle tells us something about the spin-spin force. But measuring how much a triangle one fermi across deviates from equilateral is not a simple task. However, there are methods.

If we take a distorted quark triangle model of a neutron and let it act like a free particle, it will spin and tumble about its center of mass. As a result, instead of seeing a rigid triangle, we would see a spherical smear of quarks that tends to have the up quark in the middle and the down quarks around the outside. Because the electrical charges of the two flavors of quarks are different, the outside of the neutron will tend to have a negative charge and the inside will tend to be positive. Still, the total charge will be zero—because this is a neutron, after all.

To measure this triangle we will scatter electrons off neutrons. The way electrons scatter depends upon how deep into the neutron the electron penetrates and how much charge the electron encounters. On the macroscopic, laboratory level, this translates into the probability of losing some energy or momentum. This is of course part of the "cross section" that was introduced in chapter 4.

Is there anything new here? Scattering electrons is exactly what Hofstadter was doing in the 1950s, to establish that the proton is not pointlike but a particle about 0.8 fermi across. What is new and different is that the scattering of electrons off neutrons can tell us a lot about this "distorted quark triangle." The key to this is that if the quarks are in an equilateral triangle then all the quarks are, on average, the same distance from the center of the nucleon. That means that at any distance from the center of the nucleon, the charge we would see is zero; the up quark's charge of +2/3 is canceled by the two down quarks' charge of −1/3. However, if the quarks are arranged in a line, with the up quark in the middle, the center of the neutron will have a greater positive charge and the outside will have a greater negative charge.

In the proton, the arguments are the same, but the charges and results are different. The proton is made up of two up quarks, each with a charge of +2/3, and one down quark with a charge of −1/3, for a total charge of one. If the proton's quark configuration was that of an equilateral triangle, the up and down quarks would be on top of each other, and the positive charges of the up quarks would dominate the negative charge of the down quark everywhere. Therefore, the charge distribution would be positive throughout the proton. However, we expect the proton's three-quark triangle to be distorted in exactly the same way as the neutron's triangle, except with the down quark near the center and the two up quarks nearer to the outside of the proton. But the charge of the two up quarks is 4 times larger than the single down quark's, and so the positive charges will always dominate the total charge distribution even though the quarks are in this distorted configuration. The charge distribution in this distorted triangle will rise faster on the outside and fall faster on the inside, but that is essentially indistinguishable from the proton just having a slightly larger radius (see figure 7.2). The proton does not have a strong "signature" for a distorted triangle. This is the same as saying that we cannot see the spin-dependent effects on the shape of the proton. However, if the neutron's charge distribution ever deviates from zero, it is a clear signature, a "smoking gun," of the spin effects. The realization of this signature has meant that the measurement of the charge distribution of the neutron is the hottest topic and the most sought-after experiment in nucleon structure this decade. But it is not a simple experiment and has a number of complications.

Spin will be at the heart of this experiment. It will cause the complications and resolve the problems. Spin cannot only change the mass of the three-quark system and break the symmetry of the three quarks in an equilateral triangle, but spin can also change how the electron scatters. So far we have only discussed how an electron can scatter from a charge distribution. It can also scatter from the "spin distribution" of the neutron in a "magnetlike" interaction. To see the effect of this spin interaction, first we must construct the spin distribution of the neutron in the same way we constructed the charge distribution from a

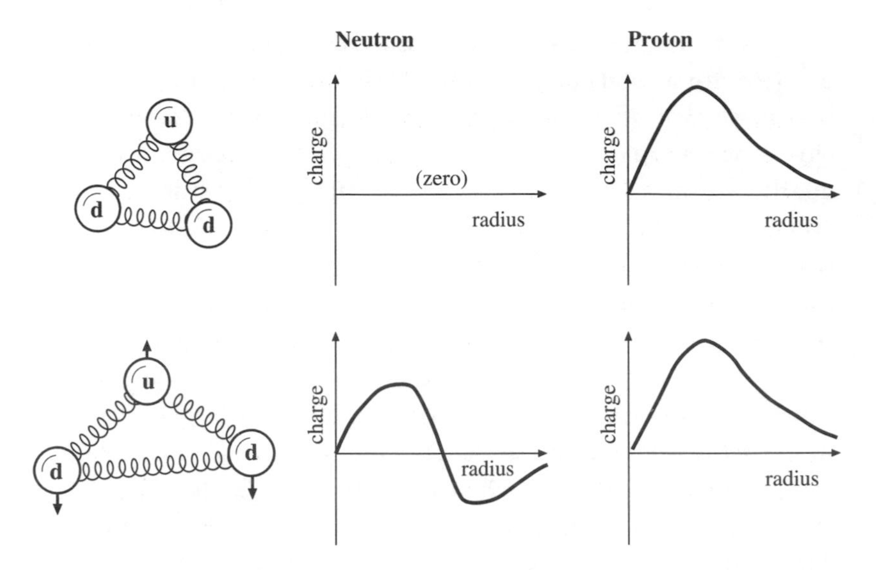

Figure 7.2 The charge distribution within a neutron and a proton depends on whether the arrangement of the quarks is an equilateral triangle or a line. The difference for the neutron is far more distinct than for the proton.

simple constituent quark model. Let us imagine that the spin of the up quark is oriented toward the left (I am picking the spin axis as something besides "up" and "down" so we don't confuse it with flavor—and because spin can be oriented in any direction). If the up quark's spin is $+1/2$ to the left, then the spin of the two down quarks—which are the same due to the Pauli exclusion principle and color—is $-1/2$ to the left. The spin distribution is "negative-left" on the outside and "positive-left" on the inside. Roughly speaking, it has the same general shape as the charge distribution. But the details are not quite so simple.

Although the charge in the neutron is the exclusive property of the three quarks, the spin is not. Gluons carry an appreciable fraction of the neutron's spin. Although gluons are fleeting, merely transitory particles that pass between quarks with the briefest of lifespans, they have an intrinsic spin of 1! On the scales where we see the effect of gluons, the difference between the charge distribution and the spin distribution should be significant. For this and other reasons, we treat the spin distribution and the charge distribution as two tangled but independent observables. This leaves us with the puzzle of how to untangle these two quantities from the data in our scattering experiment.

The key to untangling charge scattering and spin scattering is polarization. We may not know what the magnitude of the spin-dependent scattering is, or how this scattering depends on the momentum of the electrons, but we do know something about the sign of the scattering force. If the spins of the electron and the neutron are parallel, they will repel—in just the same way aligned magnets with north pole to north pole and south pole to south pole will repel. Also, like the magnets, if the spins are antiparallel, they will attract. In fact, the force of attraction has exactly the same magnitude as the repulsion, just the opposite sign, and that is the "hook" that will allow us to untangle the spin and charge scattering.

Experimentally we will make a cross section measurement, with the spin of the electrons in the beam oriented parallel to the spin of the target neutrons. We can then flip the spin of the electrons or of the neutrons and compare the two cross section measurements. By adding the two cross sections, we effectively cancel the spin-dependent part of scattering, and are left with only the charge-scattering cross section. By taking the differences of the two, we recover only the spin-dependent cross section.

Now that we have cleared the theoretical barriers, we can turn to the experimental problems—which are downright daunting. We need to orient the spins of the beam electrons, and of the target neutrons, scatter them at a precise energy, and then detect and collect all the residual particles. When we talk about spin on a macroscopic scale, as in a beam or a target, we talk about *polarization*. In one gram of matter we have roughly 6×10^{26}—600 septillion—nucleons. To get them all to march together, to align their spins in the same direction, is nearly impossible. Instead, we talk about the statistical quantity "polarization." Polarization is the difference between the number with spin-up and the number with spin-down (or spin-left and spin-right). A good polarization might be 40 percent, which means 70 percent have spin up and 30 percent have their spin oriented down.

There are several methods of producing polarized targets, but we can divide these roughly into two general categories. There are high-density targets that have impurities, and low-density targets that are very pure. With a high-density target, each electron in the beam has a greater chance of scattering off a nucleon as it passes through the target. This means that the scattering rate is high and the statistics in the measurement are good—but there are the impurities. If you have mixed the target gas with something else that will allow it to be denser, at the end you must somehow subtract the effect of that second material. This will inevitably add to the uncertainties in the final results.

The alternative approach is to work with a thin gas, a low-density target. In these targets the chances of an electron scattering are low, but if it does scatter, it will certainly have scattered off the intended target gas and not the stabilizing impurity. The inherently low statistics can be compensated for by either increasing the number of electrons, or collecting scattered particles from a large range of directions. This is the approach that has driven the rebuilding and the new experimental program at Bates Laboratory.

North of Boston, about 20 miles, poised just below the summit of a glacial moraine, is the William Bates Linear Accelerator Center of the Massachusetts Institute of Technology—or simply MIT-Bates. Bates Lab is significantly smaller and older than Jefferson Lab, but it can lay claim to having trained a substantial fraction of the physicists now working at Jefferson. In the 1980s, as Jefferson Lab was being designed and built, people at Bates recognized the need to identify a unique and important niche in the field of electron scattering. They knew they couldn't compete head-on with the newer, larger, and more powerful Jefferson Lab. The Bates community identified polarized physics as their niche, and set about rebuilding the laboratory toward this end.

Bates was originally built in the early 1970s as a 250 MeV linear electron accelerator. In the late 1970s, more acceleration cavities were added, which increased the length of the accelerator and upped its energy to 440 MeV. Then, in 1982, the beam was "recirculated." After its initial ride through the accelerator, the beam is turned around and sent up a tunnel parallel to the accelerator and then fed back into the accelerator for a second ride. This second pass doubles the energy, raising it to a total of 880 MeV. This technique of recirculating is essentially what Jefferson Lab is doing when it passes the same beam through the same linacs five times. The beam can now be steered into one of two experimental areas or into the south hall ring.

The electron beam that comes out of the accelerator is a fireball with a typical energy of 880 MeV ± 10 MeV. That burst is roughly 350 meters long and can be repeated at a rate of 600 per second. In this mode, as a "pulsed" beam, it is on only about 0.1 percent of the time. The south hall ring can change that. The ring is a beam pipe shaped like a racetrack with big bending magnets in all the curves. A beam pulse can be fed into the ring and wrapped around it twice. At the speed of light, an electron will travel around the ring one and a half million times in a second. The ring works because there are magnets around it that keep turning the electrons to the left, around and around the ring in an "orbit." If left to itself the electrons will radiate away energy every time they turn, eventually spiraling into the beam pipe, where they are stopped. So in the ring there is one, very precise, acceleration cavity. If an electron has radiated away some energy, it will arrive at the cavity out of sync with the "ideal" electron and the cavity will push it to catch up with the rest. Also, when electrons lose their energy they tend to lose their momentum in the direction perpendicular to the beam pipe. The restoring energy from the cavity is parallel to the beam pipe. This is known as "cooling" the beam—the taking out of the relative motion of electrons with the beam, and leaving only the consistent motion of electrons along the beam. The beam in effect becomes monoenergetic to better than one part in a million, three orders of magnitude better than the accelerator initially produced.

The ring has two different modes of operation: the extraction mode and the storage mode. In the extraction mode, after the ring is cooled, an electrostatic "knife" is used to peel off the outside edge of the beam. This peeling is similar to how an old-fashioned mechanical apple peeler worked. The apple was turned and a peeling knife was pushed up against the skin of the apple, which would slice off a long and continuous peel. In the ring the continuous peeling is an electron beam that can then be delivered to an experiment. It doesn't have the intensity of the original fireball pulse, but it is monoenergetic and on target essentially 100 percent of the time.

The other way to operate the ring is in the storage mode. In this mode the experiment is performed in the ring itself. Electrons in the beam can zip around the ring one and a half million times each second for up to a dozen minutes! In fact, we have had electrons in the ring for over twenty minutes. In that time they have traveled a distance equal to the distance across the orbit of the Earth, yet they are still in Bates Lab. If we put a target in the ring, an electron will pass through it on average nearly half a billion times! This means that the current of the beam has been effectively increased by several orders of magnitude. More important, the chance that an electron scatters is increased by the number of times the electron passes through the target. Actually, since the accelerator could have produced a fireball 600 times a second, the effective current increase is not the factor of half a billion, but rather something closer to a few thousand. Still, since we have chosen to work with highly pure polarized targets, which are inherently thin, factors of 1,000 are very important. The statistical uncertainties in a measurement can only be lowered with more data. This means that a high-statistics measurement that can be performed in the ring in three months would have taken 500 years without the recycled electrons in the ring.

A favorite target to place in the ring is helium-3. The most common helium, the type that fills balloons, is helium-4, which has two electrons, two protons, and two neutrons. In naming isotopes, the "helium" tells us it has the chemical properties of helium, and since chemistry arises from electrons, we know that all helium isotopes must have two electrons. For the atom to be neutral, it then follows that the atom must also have two protons. But the number of neutrons is, as of yet, unspecified. It is the number, the "4," which tells us the number of nucleons. If two are protons, the other two in helium-4 must be neutrons. In the case of helium-3, it has two protons but only one neutron—which makes it a highly intriguing target. The two protons are indistinguishable in every respect. There is no unique "quantum-color" as there is for the two down quarks inside a neutron. Therefore, because of the Pauli exclusion principle, the two protons must have opposite spins. The spins of the two protons cancel each other, as do the spins of the two electrons. This means that the spin of the nucleus, even the spin of the whole helium-3 atom, is the same as the spin of the lone neutron. Therefore, if you polarize a helium-3 atom you have, in effect, polarized the neutron.

We can group all the techniques for polarizing a gas into two different areas. The first is based on the method that Stern and Gerlach used in the early 1920s to demonstrate the existence of atomic spin. In the Stern–Gerlach case silver was heated in a oven, and the atoms that were boiled off were formed or collimated into an atomic beam. The beam of silver atoms then passed through an inhomogeneous magnetic field. Since the atoms are neutral, they should be unaffected by the field. But one of the curious properties of spin is that it interacts with the changes or "gradients" of a field. The magnetic field in a Stern–Gerlach device is designed to have strong gradients, so the flight paths of silver atoms with spin up are quite different from the flight paths of atoms with spin down. Stern and Gerlach observed that their beam split into two components on the basis of spin. We can use essentially the same technique to produce a beam of either polarized hydrogen or deuterium (deuterium is an isotope of hydrogen with an extra neutron). But a beam of polarized atoms is not a thick target. With a density of 10^{14} atoms per cubic centimeter, this target is a trillion times thinner and more ethereal than air.

The second technique involves shining an intense laser beam on the target gas. The laser light is polarized, and polarization is one of those quantities that must be conserved. So when a photon from the laser strikes an atom, it can give up its polarization by transferring it to the atom. If you dump enough polarized light into a gas, the gas will eventually polarize. Of course, polarization is like heat: it can go from light to the gas, but it can also be transferred to the walls of whatever is holding the gas. The art of producing polarized gas is the art of transferring from the laser light to gas easily, but from the gas to the walls slowly.

In all polarized targets the direction of polarization is eventually determined by a uniform magnetic field, typically produced by large Helmholtz coils. Curiously enough, the magnetic technology of polarized nuclear targets is directly tied to MRI—"magnetic resonance imaging." MRI used to be called NMR—"nuclear magnetic resonance"—because it is built on the principle that the spin of nucleons responds to a magnetic field in a well-understood and unique way. In fact, the large loops of an MRI machine are Helmholtz coils, and the method for measuring the polarization of a target is MRI/NMR. MRI and polarized nuclear targets are so intrinsically related that many of the people involved in target development are also involved in the development of higher-resolution and simpler MRI systems.

Polarized targets, both Stern–Gerlach and laser-driven, are large and complex devices. They involve plumbing for the gases, cooling, lasers, magnets, and numerous electronics to measure the polarization. They involve vacuum pumps, computer controls, and remote data-taking on just the status of the target. These targets take on a life of their own, with entire research groups that specialize in them. A modern high-quality target can be more complex than a whole experiment of a decade ago!

Finally, after choosing a polarized target and developing a polarized beam in a storage ring, we need a detector. The Bates community recognized that they needed a detector that would

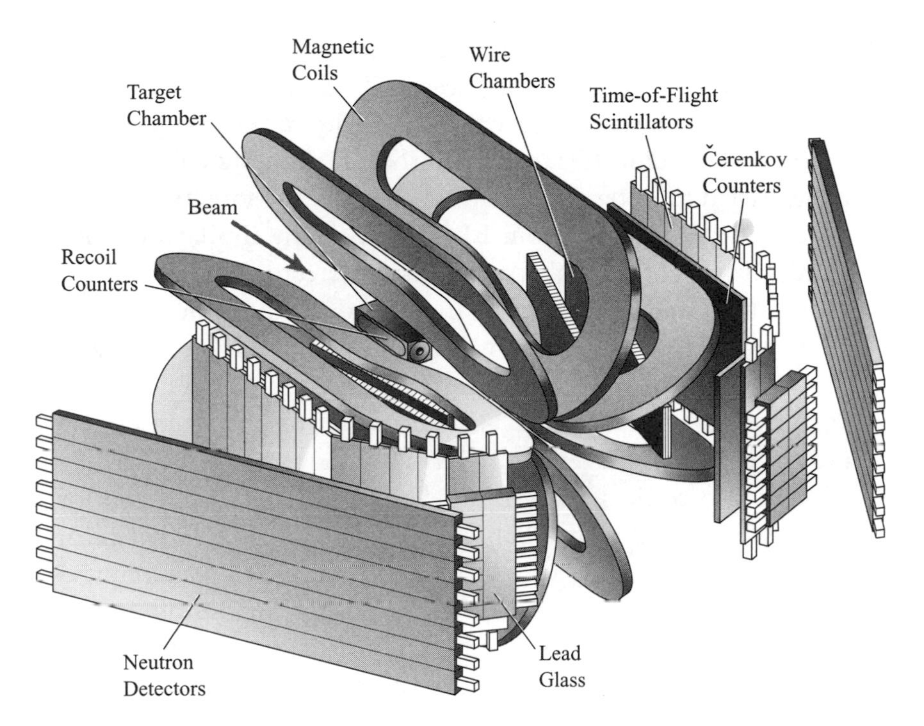

Figure 7.3 The Bates Large Acceptance Spectrometer Toroid (BLAST). Eight large magnet coils create a toroidal field around the beam. The polarized targets are built in the "doughnut's hole." Surrounding the target are layers of detectors, including wire chambers, Čerenkov counters, scintillators, neutron detectors, and lead glass calorimeters.

detect particles flying off in any direction, and so what they needed was a large acceptance spectrometer, similar to CLAS at Jefferson Lab. However, the Bates detector would have a few special features, since this is an experimental program driven by the polarized targets. The detector would be designed to fit around the targets, rather than the targets being trimmed and compressed to fit into the detector. And so the Bates Large Acceptance Spectrometer Toroid—BLAST was born (see figure 7.3). The toroidal field itself is key to BLAST. A toroidal field is shaped like a doughnut, with a hole in the center. Both CLAS and BLAST have toroidal fields, but in BLAST the hole is much smoother and larger so it can accommodate these complex and finicky polarized targets. In the heart of BLAST is a region a meter across, which is essentially free from the fields of the detector. This region will contain the targets and their Helmholtz coils, as well as the varied infrastructure that will support the targets.

Beyond the target region, BLAST has many similarities with CLAS. It has wire chambers, Čerenkov detectors, time-of-flight scintillators, and calorimeters. In addition, it has silicon strip/solid state detectors only a few centimeters from the target, and neutron detectors far from its heart. Physically, BLAST is much smaller than CLAS, but because of its ties to polarized targets and the ring of polarized beam, it is a unique and valuable facility.

I will also confess that BLAST has a particular attraction for me: I work at MIT-Bates Lab, as the software "czar." The term "czar" is used in many laboratories to describe the person who leads the software development effort. It is a curious phenomenon that the head of the part of the detector that has no budget ends up with the most ominous of titles. Perhaps this is because the developer of wire chambers or scintillators can direct that project by the way he or she buys hardware and components and hires technicians. The authority and the money follow each other. Software, on the other hand, has traditionally been left to a volunteer effort—almost an afterthought. Some desperate graduate student who wants to make some sense out of the data will develop software in the wee hours before dawn. But that model of software development doesn't work for a complex detector such as BLAST or CLAS. The volunteer element is still by and large present. Senior physicists "volunteer" their students and post-docs to write software, but this can create a disjointed "team." The head of the early software development for CLAS described this effort as "herding cats." A software czar has a title and a "bully pulpit" from which he or she can explain problems, suggest solutions, and try to make the path to a final and useful analysis code that is also the path most easy for the cats to follow. It has at times been difficult to get volunteers to write our simulation and analysis software. But the people who have stepped forward, students and post-docs from across the country and even overseas, have all been wonderful. They have made the project enjoyable and interesting and have pushed the computational envelope into regions beyond my original expectations. I would not trade my "herd of cats" for the largest budget in detector-building.

In all fairness, I must close this chapter by reporting the data we presently have on the charge distribution, and therefore the quark configuration inside a neutron. There are a number of experiments worldwide that have made, and will continue to make, measurements of the charge distribution before BLAST is completed in 2002. Still, the resolution available from a dedicated polarization facility such as the Bates Lab and BLAST will be an important contribution to the field.

Before presenting the world's raw data, there remains one last thing to explain. Whereas a theorist would like to talk about a charge distribution, and the pictures we draw of the neutron are based on that distribution, the experimentalist will talk about G_E^n (read "G E n"), the "electric form factor of the neutron." The form factor is essentially the Fourier transform of the charge distribution into a momentum distribution. The reason we talk about momentum is because that is what an experiment can measure, and because a form factor is part of the cross section we discussed earlier in this book, in chapter 4. So, finally, the world's data on the neutron form factor are shown in figure 7.4.

Physically, what does this tell us? An electron that is scattered off a neutron will lose some momentum in the collision. The momentum lost is what is plotted on the horizontal axis. So an electron that loses no momentum, which essentially misses the neutron, will "see" a neutral neutron. That is shown in figure 7.4 by the form factor being zero where the momentum is zero, at the origin. At the other extreme, if an electron has lost a great deal of momentum, it has penetrated to the heart of the neutron, and it also sees no charge. In between, the rising form factor tells us about negative charge near the outside of the neutron, and the falling form factor tells us about positive charge on the inside.

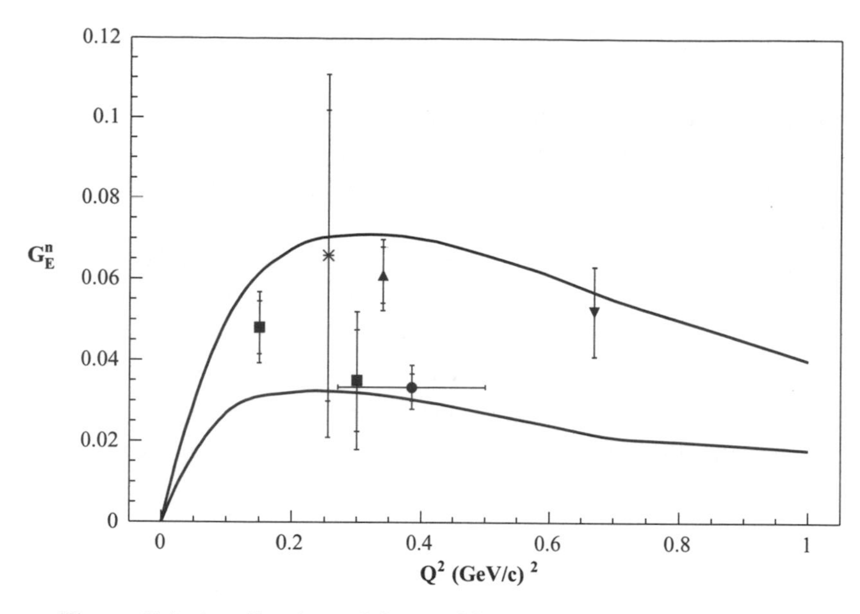

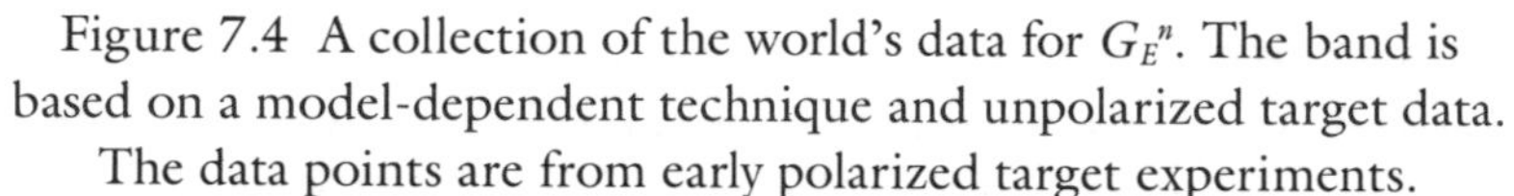

Figure 7.4 A collection of the world's data for $G_E{}^n$. The band is based on a model-dependent technique and unpolarized target data. The data points are from early polarized target experiments.

In fact, we can transform a curve that fits through the above data into a charge distribution (figure 7.5). The reason that the charge distribution is represented as a band instead of a thin line is that there is a great deal of uncertainty in the original world's data, and a range of curves might best describe these data. This is as far as an experimentalist can push the process, and now the theorists and their models must step forward and push our picture building to its end.

So far, I have always described the charge distribution in terms of the quark configuration, but, given the last curve, another explanation offers itself, especially with such large uncertainties. We could describe the neutron as a proton with a negative pion orbiting it! In fact, this is an old model of the neutron that predates the quark model. Given the present state of the data, it is not a model we can completely disregard. But I suspect it is like our three descriptions of the nuclear force in chapter 3: it is an

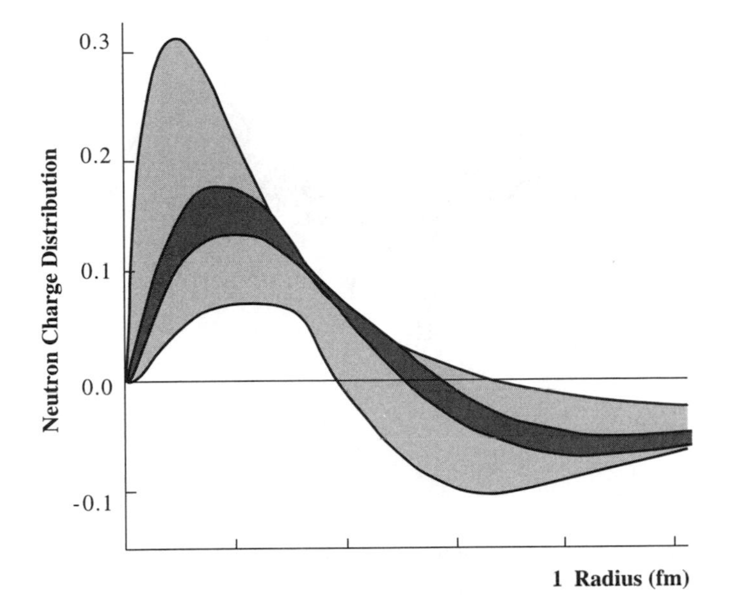

Figure 7.5 The neutron charge distribution based on a fit to all the world's data. The width of the light band is a description of the uncertainty of our present measurements. The dark band is the expected uncertainty in the next generation of experiments.

alternative description that depends on the resolution of the experiment that is looking at the neutron.

So, finally, if we read the charge distribution curve in figure 7.5 in terms of three quarks in a distorted triangle, we will see the up quark at about 0.3 to 0.5 fermi from the center, and the two down quarks at about 0.7 to 0.9 fermi. Neutrons are not rigid objects, and quantum mechanics would be talking about these distances and angles in terms of wavefunctions and probability distributions. On top of that are the great uncertainties in our measurements, which leave us, at best, with a rather gray and misty image of the neutron. Still, it is an image that will continue to come into focus with future measurements.

I close this chapter with the drawing of my best guess as to what a neutron looks like in figure 7.6. Do I really believe this picture? Yes and no. I think of the neutron really as a distorted

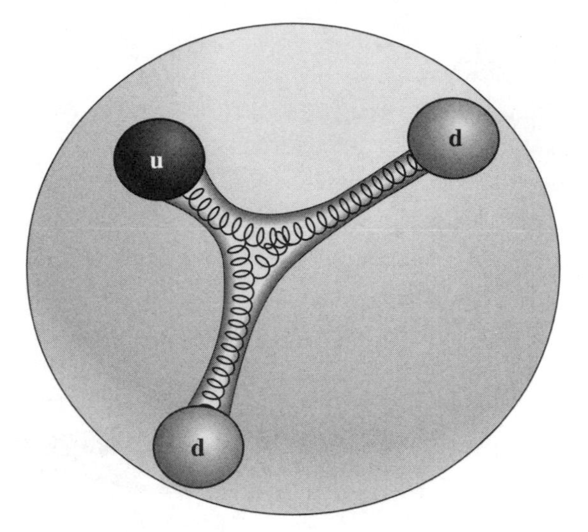

Figure 7.6 A sketch of what the three quarks might look like in a neutron—a distorted triangle.

triangle, but I have no idea how to draw the constituent quarks or the gluons realistically. Also, the neutron has a number of aspects we have yet to touch on. There are excited states of the neutron and the proton, much like the excited states of the atom. In addition, occasionally there are virtual quark-antiquark pairs: an up-antiup pair, or even a strange-antistrange pair of quarks that can exist for but a fleeting moment. But these are the stories of the following chapters.

- 8 -

Three Quarks Plus

THE QUARK MODEL has stood up well to every test we have thrown at it so far. It has explained a plethora of observed particles, everything from protons, neutrons, and pions, to the more exotic Δ, and the massive J/Ψ and Υ. In addition, it has also successfully predicted the geometry and structure of the neutron and the proton. But in retrospect, although these achievements were difficult, they were not startling. The methods we used had all been developed and refined on the much larger scale of atomic physics. In chapter 6, where we built our periodic table of the particles, we simply summed the quarks and their properties to obtain macroscopic particles with particular characteristics. In chapter 7, we essentially solved Schrödinger's equation, now with three quarks instead of the electron and the nucleus used in atomic physics. The applications may be different, but the mathematics and the technique are the same.

In this chapter we turn to two phenomena of a very different nature: the Roper resonance, and the dibaryon. The Roper resonance is a well-known resonance that has been measured repeatedly. We might be able to explain this resonance in terms of an excited state of a nucleon, a nucleon with one quark bumped up into a higher orbit in the same way electrons can be bumped up into a higher orbit in an atom. But there is also a second explanation, which involves gluons, for which there is no parallel in atomic physics.

The dibaryon is, as of yet, only a theoretically predicted particle built up of six quarks. Dibaryons seem to be allowed by QCD and the quark model, but they have never been observed. If a dibaryon is ever found, it will be exciting, since it is truly a new form of matter. However, if it is not found, we must recognize that the rules for color combination—that only three quarks can combine to make "white"—is not just the normal state of affairs, but an absolute and truly fundamental law of nature.

So in this chapter we push the quark model beyond its original scope and apply it to phenomena unconceived of when the quark model was originally designed. The exercise of pushing a theory or model has often caused it to crumble. But this is the crux of the scientific method, the crucible and fire in which we assay the theory. If the theory fails the test, we either discard the theory, recognize its limits, or correct its errors. As a classic example, Descartes proposed three "laws of motion" in about 1632. Clearly René Descartes was a gifted thinker and his mechanics should not be scoffed at. He set out to explain collisions and quickly identified momentum as a key element in this problem. In some respects, the introduction of momentum as a new concept was an intellectual feat in itself. Momentum is a constructed quantity. It is neither length nor mass nor time, but a function of all three. It is something you cannot see or feel, a quantity that is not visible without an interaction. There was no precedent for such an abstract concept. So when Descartes penned his laws of motion he started with his two "laws of persistence": an object in motion stays in motion, and the trajectory of an object will not curve unless something else pushes on it. These two laws are essentially rolled into Newton's first law, or "law of inertia." Descartes then crowns his treatise with his "law of impact," which we would recognize as simply the "conservation of momentum." The conservation of momentum is a good and valid law. In fact, a quantity that obeys a conservation law is technically referred to as a "canonical momentum," which forms the basis of any good system of "mechanics." The word "canon" itself emphasizes the role of universal or divinely ordained. For example, there is a conservation of energy law because energy is a canonical momentum. But in Descartes's case the momentum law was very limiting. His mechanics starts and stops with the collision of various billiard-ball-like objects—but went no further.

Newton's laws of motion, published in *Principia* in 1688, are only slightly different on the surface. They differ primarily in identifying "force" as the critical element. In fact, Newton's laws are so tightly associated with force that we commonly say, "according to Newton's laws," and then only cite one, "$F = ma$." This second law is the only one unique to the mathematician from Cambridge. It is the second law that gives dynamics to Newton's equations and made Newtonian mechanics the core of the physics curriculum for more than 300 years. And it is this mechanics that serves as the prototype for quantum mechanics and for any other mechanics.

But why do Newton's equations eclipse Descartes's, Hobbes's, and others'? The reason is that they explained every phenomenon that earlier mechanical systems had described, such as collisions and simple motion—and then added the arch of a cannonball and the planets in orbit. Newton's equations accurately describe toy tops that don't fall down and the angular momentum of tumbling satellites. For hundreds of years every new phenomenon that was thrown at Newtonian mechanics was successfully analyzed and explained. Every new challenge reinforced our faith in these three laws and demonstrated their robustness and versatility.

I don't expect the quark model, or even quantum chromodynamics, to be the last word on the most elementary of particles; string theory and other "grand unified theories" (GUTs) and "theories of everything" (TOEs) are knocking on the door. But the quark model and QCD seem to be well enough founded that they should be able to explain phenomena beyond those that originally inspired them, and which we have already discussed. They should be able to take a little "heat."

Our everyday nucleon has a mass of about 940 MeV/c^2. As we raise its temperature, by pushing more and more energy into it with either electron bombardment or showers of virtual photons, nothing happens. The nucleon remains inert and inanimate until we dump about 300 MeV of energy into it. At that point it can absorb the energy—albeit briefly—by flipping the spin of one of the quarks and transforming itself into the now familiar Δ particle. Spin flipping cannot store any more energy, because there really are only two unique spin configurations: all parallel, or one antiparallel. Still, we are curious to see what happens when we raise the temperature even higher.

In the mid-1960s, this is what was happening at various laboratories throughout the world. The accelerators were new, the detectors were unfamiliar, and the physics was mysterious. It fell to a young physicist, L. David Roper of the Lawrence Berkeley Laboratory, to paw through the data and pull out the newest of gems—a resonance at 1,440 MeV/c^2. We still use the term "Roper resonance" to refer to that first resonance. In addition, two more resonances followed close on the heels of the Roper, at 1,560 MeV/c^2 and at 1,610 MeV/c^2, as shown in figure 8.1.

When we recall that a resonance in our cross section is like a bright line in an atomic spectrum, an explanation of the Roper comes to mind. For clearly the Roper is only an excited state of the nucleon, and we should be able to draw an analogy to excited states in the atom. In atomic physics energy is released when an electron drops from an outer orbit to an inner orbit. In fact, essentially every combination of atomic orbits is related to a different line in the spectrum. For example, when a hydrogen atom is in its first excited state, it has a mass of 938,896,599 eV and the electron's orbit is about 4 Å across. When the electron drops into the ground state, the diameter of the orbit drops to 1 Å and the mass of the atom drops to 938,896,596 eV. In the process a 3 eV photon is produced, which is seen in the violet part of the spectrum.

There seems to be no reason why quarks in a nucleon should not act in the same way as electrons in atoms. The "distorted-

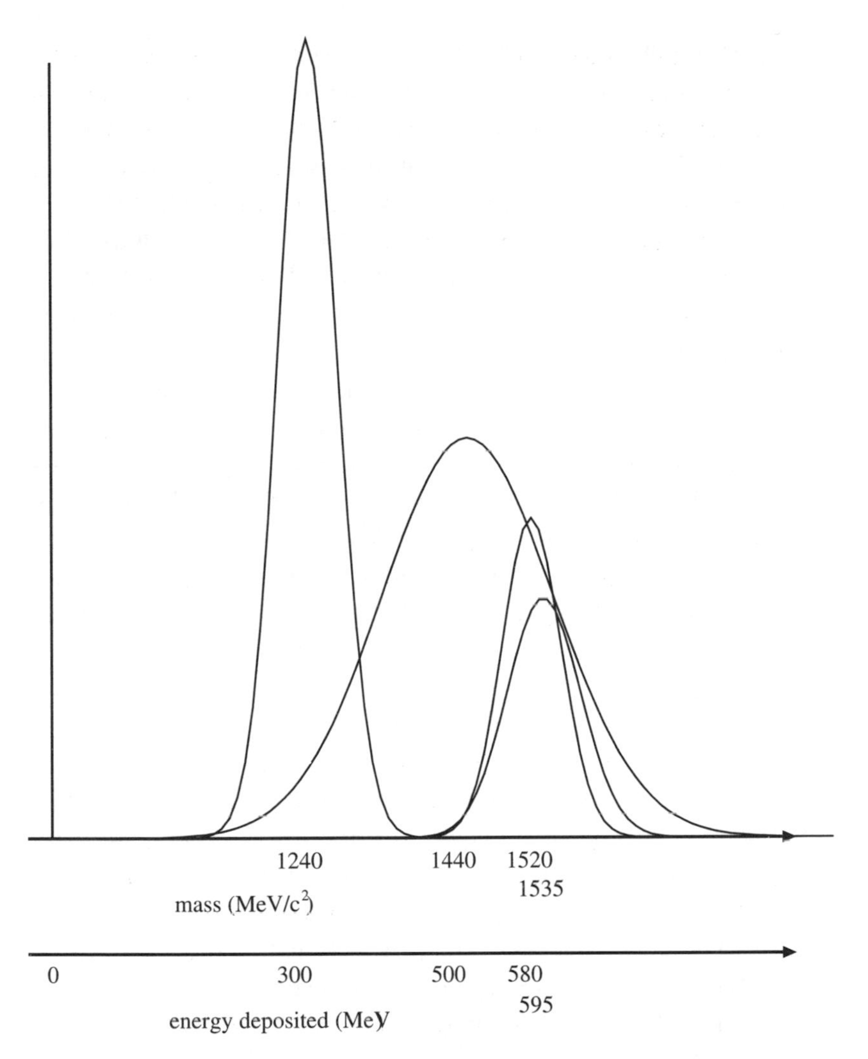

Figure 8.1 In the cross section of a nucleon we see a number of "bumps" or resonances. The first, the narrowest, and the most important is at 940 MeV/c^2, when we have added no energy to the nucleon; this is the nucleon itself. When we add about 300 MeV of energy to a nucleon, we can create a Δ particle. By adding 500 MeV, we can "ring" the nucleon. This resonance is called the Roper. The resonances of the nucleon at 1,560 and 1,610 MeV are generally referred to by their masses as N(1560) and N(1610).

triangle" configuration described in chapter 7 is the ground state of the nucleon. Basic quantum mechanics tells us that in the ground state of a system particles travel in spherically symmetric, essentially circular orbits close to and around the system's center. But they are not quite the orbits we are familiar with. We are used to a "planetary" picture, with a central star or nucleus much more massive than the orbiting "planets." The situation for three quarks is much different. The orbit is more democratic, since all three particles have essentially the same mass—although the forces among them are a complex, spin-dependent tangle.

Now "orbit" might give us a nice and compact image, but since this is a quantum mechanical system, we should be talking instead in terms of wavefunctions, or probability distributions. These wavefunctions for three particles with a complex interaction are themselves complicated beasts. The theorist who wants to compute the orbit can approach the task in one of two ways. First, she can simplify the problem by describing the interaction between quarks as "springlike," and then proceed with the rules of calculus to derive a wavefunction. Alternatively, she might use a more realistic interaction derived from QCD and rely on computers and numerical methods to grind out a solution. In either case there are some very general results: the ground state is related to the quark distribution described in the preceding chapter, and the excited states are larger and have more energy. But before proceeding it is probably worthwhile to pause and discuss just what exactly is meant by an "excited state."

How is it that one system can have many different "states"? Actually, that is not so unusual, for it is no different than saying that one equation can have many solutions. But these solutions are intimately tied to one another. Because quantum mechanics follows the mathematics of waves, we can try to understand these through some human-size wave phenomena. For what is true about the wavefunctions of atoms and quarks is also true about the waves related to light and sound.

What happens, for instance, when you pluck the string of a guitar? A wave is set up in the string that vibrates with a particular pitch, or frequency, depending on the length and tension of the string. But there are also other frequencies mixed into the sound we hear, the "harmonics." The first harmonic is related to the sound we would hear if the string were half as long, and that harmonic has twice the frequency of the dominant sound. Similarly, there is an entire series of higher harmonics, pitches that would be generated by strings a third, a fourth, a fifth, and so forth, as long as the original string. In fact, it is the relative strengths of each harmonic that make a "C" note on a guitar different from a "C" on a piano, a violin, a flute, or a trumpet. Here, however, the analogy with the quantum world breaks down. A flute playing a middle-C also plays a number of harmonics at the same time. An atom or a nucleon can only play monotonic notes. Such a particle can switch from one harmonic (or orbit, or state) to another, but it can only play one harmonic or state at a time.

So a nucleon has a number of different orbits, or wavefunctions, where a quark might be found. If a nucleon is "excited," we mean that a quark has been "kicked" into a higher orbit. This language always makes it sound as if the quark were like a NASA satellite being pushed from a low, space-shuttle orbit, into a higher, geosynchronous orbit. The Earth's motion, as well as the trajectories of other satellites in space, is essentially unaffected by the kick. But for a nucleon, in which a third of its mass has just shifted, the effects are prodigious. One quark has dramatically rocked the boat. And that makes the entire question of what the excited state looks like a highly complex problem. The Pauli exclusion principle, for instance, which states that the two down quarks in a neutron must have parallel spins, holds no sway if the quarks are in different orbits.

Once we have opened Pandora's box in this way, the possibilities are endless. If we can have a first harmonic, we can also have a second, a third, and higher harmonics without limit. The states need not be spherically symmetric either; an admixture of angular momentum is allowed in these excited states. And even if there was a series of excited states, observationally their peaks in the cross section would appear to smear together, for each one is unstable, with a very short lifetime and therefore a very wide resonance.

But whatever the complex details of these excited states might be, a few signatures stand out clearly. The nucleon should be bigger and the orbits of the quarks wider. We generally describe these excitations as radial ones. And radial excitations offer a clear and conventional explanation of the Roper resonance, a simple kicking of a quark into a higher orbit.

Yet, as I promised, there is a second hypothesis that might explain the Roper, for which there is no parallel in atomic physics. The Roper resonance might be a "gluonic excitation." Until this point in the book I have avoided the issue of what a gluon is because I wanted to develop quarks fully and introduce gluons only when needed. We need them now.

Gluons are the virtual particles that carry the QCD strong force between quarks. They are the equivalent to photons in electromagnetic interactions or pions in the nuclear force. A quark is attracted to a neighboring quark because of a gluon passing between them. The simplest Feynman diagram for a quark-gluon interaction is shown below. If we also diagram the flow of color within the quarks and gluons, a most curious pattern emerges (figure 8.2). It is a pattern highly reminiscent of the "pion exchange" and "pion-as-quarks" description of the nuclear force (see chapter 3). The parallels are strong and instructive, but gluons do harbor a few unique features. First, a gluon, unlike a pion, is never seen outside boundaries delimited by quarks. This feature, of course, is related to the issue of quark confinement. Second, instead of electrical charge, gluons carry "color charge." To understand the role of color charge, consider the gluon exchange in figure 8.2. The up quark starts out red and then emits a gluon. The down quark starts out blue and absorbs that gluon. Besides absorbing the momentum of the gluon, the down quark must also absorb the color, its "redness," and emit its "blueness" backward in time. So this intermediate gluon carries a color charge

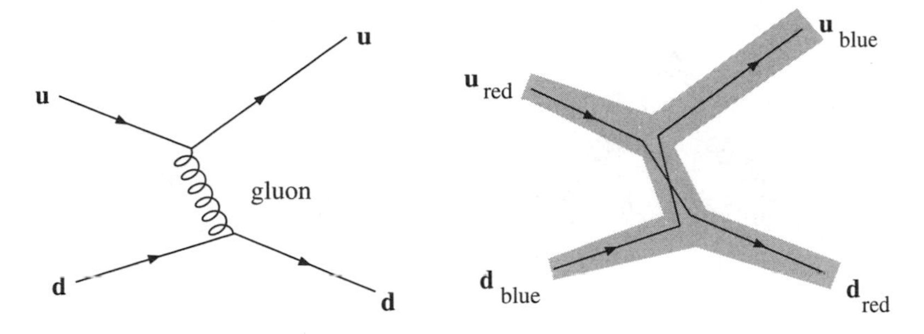

Figure 8.2 *Left:* The Feynman diagram for an up and a down quark interacting by passing a gluon between them. *Right:* If we diagram the flow of color, we see that "red" is flowing forward through the gluon, while "blue" is flowing backward through time. This means that the color of this gluon is $r\bar{b}$.

of red-antiblue. In fact, there is a whole complement of gluons, each with a different permutation of color charge.

The third unique feature of the gluons is that they readily interact with other gluons. In QED photons do not interact with photons. In fact, the gluon-gluon interaction is really a result of that second property, the fact that they carry a charge. But this feature is so important in the pictures we draw of quarks and gluons that it is worth noting in its own right. For example, I now can have two different three-quark three-gluon pictures of a nucleon (see figure 8.3). Self-interaction by itself is just a curiosity, and a complication for the theorist to unravel. But when we pair it with our last property, "strong coupling," the nucleon is ready to assume a new dynamic, a new attribute.

We have always called the QCD force the strong force—since the force that holds a nucleus or a nucleon together is much greater than the force that holds together an atom. But "strong coupling" is different. The magnitude of interaction in quantum electrodynamics is determined by a number called "α," which is roughly equal to 1/137, or about 0.0073. In the mathematics of QED, the interaction of one photon and an electron is proportional to α. If there are two photons and two interactions, the

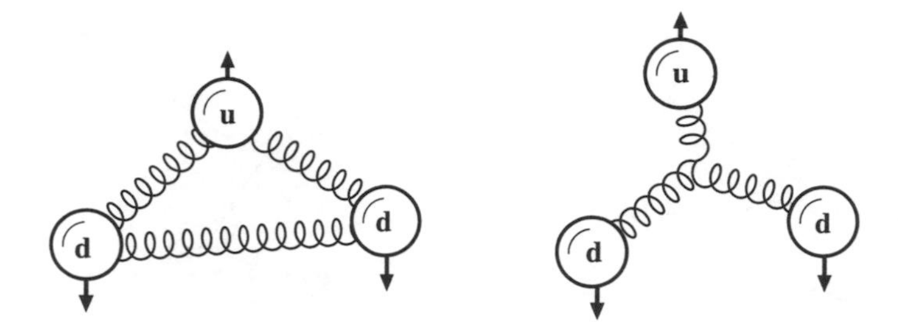

Figure 8.3 Two different pictures of a three-quark/three-gluon system. The second figure has no parallel with photons, for, unlike photons, gluons can interact with gluons.

scale factor is proportional to α^2, or about 0.000053. It is true that there are a number of different two-photon contributions, but still, the one-photon interaction is 100 times more important than the two-photon interactions. The same is decidedly not true in QCD.

In QCD the coupling constant is called α_s. Actually, the value of α_s depends on the scale of the physics being explored, but in our discussions it has a magnitude of roughly 0.1 or so. Since it is so much larger than α, and because the number of different interactions increases exponentially with the number of gluons, we cannot ignore the more complex diagrams. In fact, although the single-gluon interaction is still the most important, even the three-gluon interactions, which are proportional to α_s^3, have a significant effect! Now putting the mathematics aside and returning to our task of just drawing a picture of the nucleon, we really must include a web of gluons (see figure 8.4).

Now with a web of gluons in our thoughts it is not a great leap to recognize that this web may be able to exhibit its own dynamics. Although this view of gluons acting as a web or mesh doesn't lend itself readily to calculations, it does open the door to new possibilities. For the purposes of explaining the Roper, there is the very real potential for the gluon mesh to exhibit its own vibrational modes! And of course a new vibrational mode would be a way in which energy could be temporarily stored, and a mechanism by which we might explain the Roper resonance.

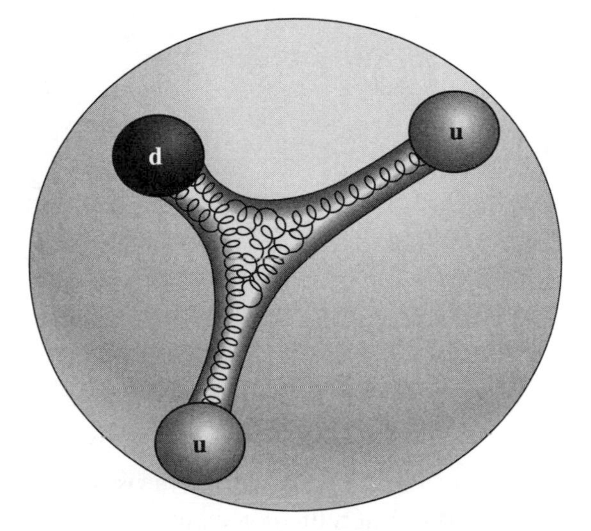

Figure 8.4 Since gluons interact with each other so readily, our picture of a nucleon should include a tangle or web of gluons.

To recap this discussion of the Roper resonance, it is useful to step back and look at what a resonance is. In the most general sense, a resonance is a way that a system can rearrange some of its internal components to store energy briefly, and then release that energy. In the case of the Δ particle, a spin flip stores the energy for about the same amount of time it takes light to cross two protons. The excess energy in the Δ is then released by the creation and ejection of a pion. There are two proposed mechanisms for storing energy in the Roper resonance. First, the Roper might store its energy as a radial excitation. The energy we see in the resonance is the energy released when the quark drops from the excited state to its ground state. Alternatively, a Roper might absorb its energy by "plucking" the mesh or web of gluons. The energy is then stored in vibrational modes of the gluons.

So how do we distinguish these two mechanisms? Our first thoughts are to turn to the theorist and ask, Which mechanism will produce a particle—or resonance—with a mass of 1,440 MeV/c^2? The calculations are long and difficult, and riddled with a number of uncertainties; for they depend on a knowledge

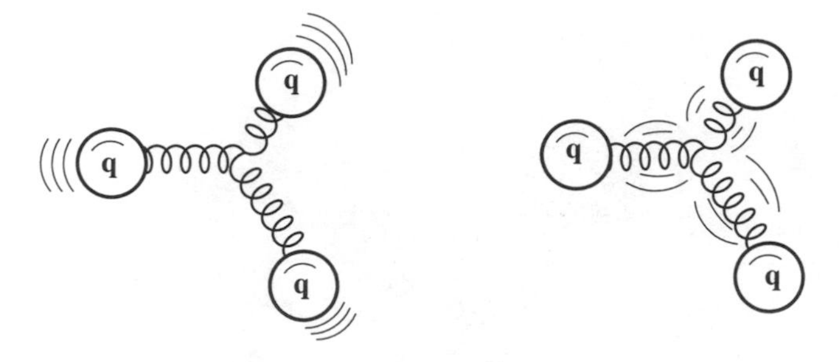

Figure 8.5 Experimentally we can excite either the gluons or the quarks. The difference will help us untangle the modes of oscillation that drive the Roper resonance.

of aspects of nature that we have yet to measure precisely. But in the end, within the inherent uncertainties of the theoretical models and calculations, *both* mechanisms can explain the Roper resonance's mass of 1,440 MeV/c^2! We should not be disheartened, though—there is still a hook on which we can hang our hopes.

When we created a Roper, did we scatter our photon off a quark or a gluon, as shown in figure 8.5? The curious properties of photons will help us find the answer. Real photons, like the ones that travel across the universe, have to be self-sustaining. That is the miracle of Maxwell's equations of electricity and magnetism. A real photon can sustain itself, essentially regenerate itself, because its modes of electric and magnetic oscillation are perpendicular, or "transverse," to its direction of motion. A virtual photon doesn't have this constraint. A virtual photon, which lives on borrowed time and momentum, doesn't have to be self-supporting, and it can exhibit "longitudinal" modes of oscillation—like sound waves supported by the medium of the air. And it is these longitudinal modes of a virtual photon that excite the gluons.

Finally, our old friend polarization returns to give us a handle on these longitudinal and transverse photon modes. By orienting the polarization of the beam and the target either parallel or perpendicular to each other, we can control the transverse and longitudinal components of the virtual photons, and therefore effectively turn off and on the Roper resonance!

So which is it? Radial or gluonic excitation? The experiments have been planned—but so far there are no definitive results. I can only add this cautionary warning: there seems to be no reason why it cannot be both. The Roper might be a hybrid of mechanisms. A hybrid will make the results less distinct. But still, embedded in that resonance would be gluonic excitation—an unparalleled phenomenon in quantum mechanics.

It is ironic that the strongest force in the universe results from the QCD color charge—yet we never directly observe anything with color. This follows from the rules of quantum chromodynamics, which require that any observable particle be "white," or color-neutral. The neutrality clause is satisfied by only two classes of particles: baryons and mesons. Baryons are three-quark combinations that achieve their "whiteness" by summing all three QCD colors: red plus blue plus green is white (try this with light, not with paint). On the other hand, mesons, which are built out of one quark and one antiquark, achieve their neutrality by combining color and anticolor. For example, a pion might be made up of a red quark and an antired-antiquark. But observationally speaking, there are no more combinations: we have never seen in nature any other solutions than the color-neutral baryons and mesons.

But why? The equations of quantum chromodynamics do not explicitly identify these two combinatoric solutions. Any combination that is "white" should suffice. The five-quark combination red-green-blue-red-antired is a theoretical solution to QCD. Likewise the six-quark combination red-green-blue-red-green-blue should satisfy the rule that calls for whiteness. Yet these alternative color combinations have never been observed.

A number of arguments about the intrinsic stability and lifetime of these unobserved combinations have been put forth to explain why they would exist only for the most fleeting of moments—and maybe not at all. But a number of detailed calculations have also been made, predicting that the dibaryon—the six-quark combination cited above—should be observable. In fact, these predictions have been around since 1977, and a number of experimental searches for the dibaryon have been conducted, all without success. As a consequence of this quarter century of disappointment, most physicists are ready to give up the search. Other physicists feverishly argue, however, that the search strikes at the core of QCD and the allowed color combinations. If there is something beyond baryons and mesons, we should be looking for it. In some sense it is as important as finding the top quark. But also, if there really are no dibaryons, that too is of fundamental importance. These zealous dibaryon hunters also argue that only with the most recent generation of laboratories and detectors can we hope to be able to find these elusive particles.

What would a six-quark particle look like, and how could we distinguish it from the familiar deuteron? A deuteron is made up of a neutron and a proton, which means that at the quark level it is built up of three up quarks and three down quarks. In fact, deuterons are fairly common, forming the nucleus of a naturally occurring isotope of hydrogen. But the deuteron is not the six-quark object we are looking for. Back in chapter 2 we described the nuclear force between protons and neutrons in terms of pion exchange, quark exchange, or even a color Van der Waals force. In all of these descriptions three quarks are grouped together into physically localized and color-neutral nucleons. These nucleons never directly exchange gluons, and there is a clear division about which quarks belong in which nucleon.

A dibaryon, if it exists, is a very different beast. The gluons would be exchanged among all six quarks in a most homogeneous manner, and physically all six quarks would occupy the same region. But it is very tricky to determine experimentally which quarks exchange gluons. An experiment that could be a shade simpler to carry out is to measure the size of the particle; the dibaryon should be significantly smaller than the deuteron. In fact, the size measurement is often cited as the litmus test that could distinguish a dibaryon from what is merely an unusual variation on a deuteron. Yet talking about measuring the size of a potential dibaryon is getting way ahead of ourselves—for at this time we don't even have a dibaryon candidate.

So where should an experimentalist hunt for the dibaryon? The theorist can offer some guidance that will at least narrow the range of resonances that would have to be examined. First, let us turn to that well-studied and highly stable six-quark system: the deuteron. At 1,876 MeV/c^2, the deuteron sets the lower limit on the mass of a dibaryon. If the dibaryon were lighter than the deuteron, the deuteron would eventually decay into it. But the deuteron is exceptionally stable. This raises the natural question, If the dibaryon is heaver than the deuteron, why doesn't it immediately decay into a deuteron? This question seems much like an earlier one: why doesn't a Δ particle instantly decay into a nucleon? In that case the Δ particle exists because it takes time to get rid of the extra spin. If dibaryons do exist, their stability might also be tied to their spin.

The sum of six spin $\frac{1}{2}$ quarks could be 0, 1, 2, or 3. Immediately we can eliminate the spin 1 combination, for the deuteron is spin 1. If the dibaryon were spin 1 as well, it would decay so fast into a deuteron that it would never be stable enough to call itself a particle. Its lifetime would be too brief for us to "see" any evidence of it. The theoretical argument that narrows the potential spin and mass of a dibaryon continues in this piecemeal fashion, eliminating a spin here, a mass range there. For example, the dibaryon cannot have spin 3 and a mass greater than two Δ particles, or it would decay into two Δ *s*. In fact, after considering case after case, the range of possible homes in which the dibaryon might lurk narrows to this: the dibaryon can have spin 0 or 2 and its mass must be between that of a deuteron (1,876 MeV/c^2) and that of a nucleon-Δ combination (~ 2,140 MeV/c^2).

In addition to these general considerations, there have been a number of detailed calculations predicting that the dibaryon not only will be found in this range, but will also be relatively stable, compared with the Roper or the Δ. That stability translates into a resonance in the cross section that might only be a few MeV wide.

To date, no one has ever seen a dibaryon—but recently a team of physicists working in Uppsala, Sweden, reported a new resonance at ~ 2,100 MeV/c^2 that is only 4 MeV/c^2 wide. It could be argued that a resonance only 4 MeV/c^2 wide is on the hairy edge of their experiment, and that the results are far from conclusive. But as interesting as these experimental results were in themselves, even more interesting was the way this announcement polarized the physics community. Most physicists were highly skeptical, reasoning that this energy was in old and well-trod territory, and if this resonance really existed we should have seen it by now. The other camp was excited by the possibilities. Now, with a narrow area to focus on, and a new generation of accelerators and detectors, they felt that the hunt could really start in earnest. In fact, by the year 2000 the Uppsala group had collected a great deal more and better data, and withdrew their original claim. Even now, however, the possibility of finding something beyond mesons and baryons is still just too intriguing to ignore.

Is the Roper resonance a gluonic or a radial excitation? Do dibaryons exist? Both of these questions are tied to the most fundamental aspect of QCD: gluons in the case of the Roper, and color counting rules in the case of dibaryons. Curiously enough, these most fundamental questions can be addressed with macroscopic particles and resonances. QCD can be probed and prodded on the level of nucleons and a few GeV of energy, far from the genesislike fireballs of FermiLab and CERN. For a decade or more, problems and questions have stood out in front of the experiment, beckoning them to follow. But in the next decade, with new experiments being mounted, older problems will be probed and undoubtedly new anomalies will be uncovered. For this is the way science is supposed to work: good science is a marriage of curious questions and challenging experimental problems. It is remarkable to see QCD, one of the most fundamental theories in physics, being played out in the realm of neutrons and protons—in the realm of the stuff that makes up our everyday world.

- 9 -

Digging a Little Deeper

WHEN Heinrich Schliemann first pushed his shovel into the earth at Hissarlik, in Turkey, he fully expected to uncover the ancient city of Troy. Homer's *Iliad* had conveyed to him such a vivid picture of that city of Priam and Hector, of Helen and Paris, that he felt he already knew what lay below his feet. He envisioned the Homeric fortress, the palace of Priam, the treasure houses, and the armories. What he did not envision was *nine* complete cities stacked on top of each other. Layer after layer, stratum after stratum, of distinct civilizations had built on the same hill. There, encased in the rocky soil of Hissarlik, was an exhibition of nearly 5,000 years of humanity.

What a shame it would have been if Schliemann had stopped in the basement of the uppermost layer and never found the citadel of Priam! Perhaps he didn't realize it at the time, but that uppermost layer was a flag, an indicator that this hill was intrinsically attractive to city builders; it had the right geography for a metropolis. So Schliemann continued to excavate. Where we find a fruitful area, we are well advised to examine the next stratum down: to dig a little deeper.

In our exploration of the anatomy of a proton and a neutron, examining the next stratum means turning up the magnifying power of our microscope. The constituent quarks, which have been the central focus of most of this book, are most apparent at

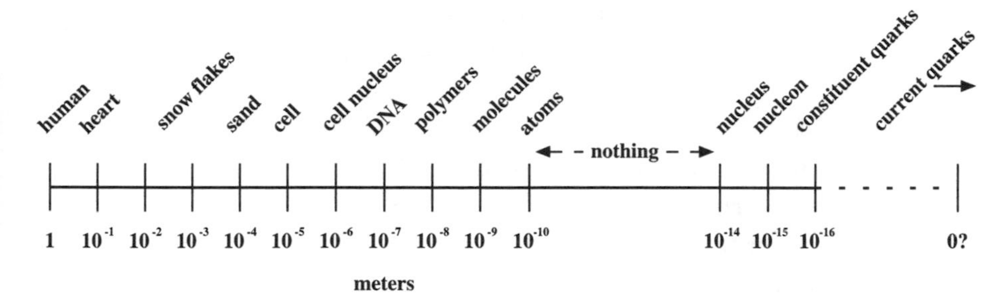

Figure 9.1 Between the scale of an atom and a nucleus are four orders of magnitude—a factor of 10,000. This region is void of any physical phenomena. Yet between the nucleus and the scale where we need current quarks to understand the structure of matter is only a factor of 100. As if to make up for lost time, nature packs four different levels of nature: nucleus, nucleons, constituent quarks, and current quarks, with current quarks being the right description down to zero.

a scale of 0.3 fermi, or about a third of the size of a nucleon. But if we increase our resolution and magnification power by another order of magnitude, the constituent quarks seem to dissolve into their most basic components: the "bare" or "current" quarks, and the gluons.

One of the things that makes the physics at the scale of the nucleons and quarks so fascinating is that within two orders of magnitude, between 10 and 0.1 fermis, we drop through four distinct strata of nature. But first let us step back to get some perspective on these strata.

If we could focus our "microscope" on any piece of matter, with a resolution of 10^{-10} meter (one angstrom), we could see atoms (see figure 9.1). With a slightly better resolution (10^{-11} meter), we could readily distinguish the electrons in their atomic orbits and the nucleus at the center. Now, if we increase our resolution by another factor of 10, to one one-hundredth the size of an atom (10^{-12} meter), we would again see just the nucleus. This nucleus would look the same whether it came from intergalactic plasma or from the heart of a human being. Again, if we in creased our resolution by another order of magnitude, to 10^{-13}

meter, we would still see a pointlike nucleus. In fact, even at 10^{-14} meter, with a magnification that enables us to resolve objects one ten-thousandth the size of an atom, we would only see a pointlike nucleus.

Only when the resolution reaches 10^{-15} meter—one fermi—can we see structure again. At this scale the nucleus can finally be resolved into a collection of protons and neutrons, with pions fluttering back and forth between them. The nucleons and the pions, like the atoms at the chemical scale or the electrons and nucleus at the atomic scale, form an essentially complete description of what is seen. From these components we can calculate orbits and energy levels, we can derive the choreography of the dance of the nucleons. But we humans are curious creatures, and we will continue to dig.

Nature now seems to make up for the scarceness of phenomena across all those scales. By increasing our resolution by a factor of only about 3, to 3×10^{-16} meter, a third the size of the proton, we find ourselves peering into the realm of the constituent quark. We then have to go no farther than another order of magnitude to find ourselves among the current or bare quark and gluons. Between 10 fermis and 0.1 fermi, we have passed from nucleus to nucleon to constituent quarks to bare quarks—four unique and internally consistent descriptions of matter. There are few other scales in nature where the void is so wide and the phenomena become so tightly packed.

Following this frenzy of phenomena and strata, we seem to have hit the experimental basement. Even if you turn up the magnification by an additional factor of 1,000—as you can at FermiLab or CERN—there appear to be no more layers of matter, no further strata. Bedrock down until "?"

So what is a current or bare quark, as opposed to the constituent quark that we have spent so much time developing? The answer has to do with both the scale and the role of virtual particles. Physicists see these quarks as being the "real" quarks, more real than the constituent quarks, because they are pointlike and therefore more fundamental. The term "current quarks" is commonly used, which emphasizes their role as something that gives rise to the more macroscopic and "almost observable" constituent quarks, in thesame way that a current in a wire can give rise to an electric and a magneticfield. The alternative adjective, "bare," emphasizes that this is the quark with our view unimpaired by the cloud of gluons that surrounds it. But we are getting ahead of ourselves. Let us just start with the statement that bare quarks are the building blocks of matter on the next smaller scale. They are the way we see nature when we increase the power of our microscope by another factor of 5 or so beyond the size of the constituent quark. What we see now is a confusion of quarks and antiquarks as well as the increased role of gluons. Yet before I can draw a picture of a nucleon at this scale, let me develop in more detail the role of virtual particles.

Virtual particles have faded in and out of our discussion ever since chapter 3, where we first encountered pions in Yukawa's theory of nuclear force. The pion connects nucleons in a nucleus and conveys momentum between them. Although these virtual pions can carry momentum and energy, and therefore have the quite real and significant effect of holding the nucleus together, they themselves are considered unreal or "virtual." They have no permanent mass or lifetime of their own. They live on borrowed time and energy, as allowed for by the Heisenberg uncertainty principle. The uncertainty principle tells us that the more mass that is borrowed, the shorter the lifetime of the loan. For example, if the nucleus creates a virtual pion with a mass of ~ 140 MeV/c^2, the pion can only exist for about 10^{-23} second, and can travel only about as far as the distance across the nucleus. But we have seen all of this already.

The twist on virtual particles that we need to add to understand current quarks is that the Heisenberg relations are limited to pairs of observables that can have continuous values, such as time and energy, or position and momentum. But these relations do not extend to quantities that have discrete values, such as spin and flavor. In fact, if we look back at our diagram of "pions as a quark exchange" from chapter 3, we notice a curious property.

At any given instant, the number of quarks minus the number of antiquarks is constant. But the pattern goes much further than that. At any given moment the number of up quarks minus the number of anti-up quarks is also constant. Likewise, spin up minus spin down is constant, and of course charge is always constant as well. These numbers, which have discrete values, are all referred to as "good quantum numbers," for they are conserved even on the microscopic level. These properties are truly quantized properties.

There are just a few rules that any particle, virtual or real, must obey to be created. First, the quantum numbers must be conserved, which means that particles are usually created in pairs. For example, a photon can decay into a quark-antiquark pair, or an electron-positron pair (a positron is an "antielectron"). The second rule governs the fate of these newly born pairs. If they were produced with enough energy to generate both of their masses, they can travel off in any direction, with real identities and separate existences. If, however, there was not enough energy to endow the particles with a real existence, they are a pair of virtual particles, living on borrowed time and energy.

In the case of quantum electrodynamics (QED) these two possible fates are quite well understood and experimentally established. If the energy of a photon is greater than 1.022 MeV, or in other words, greater than twice the mass of an electron, the photon can decay spontaneously into an electron-positron pair. Both particles can then be measured and observed in the laboratory. If, however, the energy of the photon is less than 1.022 MeV, the electron and the positron will recombine into a photon.

Are pairs of electrons and positrons created and then reabsorbed before they interact? And if so, how could we know this? Again let us look at that simpler and well-studied theory: QED. When we write down the equation for a process such as scattering, we write it as the sum of an infinite series of terms. The reason QED ranks as one of the most successful theories is that the infinite series rapidly converges. The leading term of our

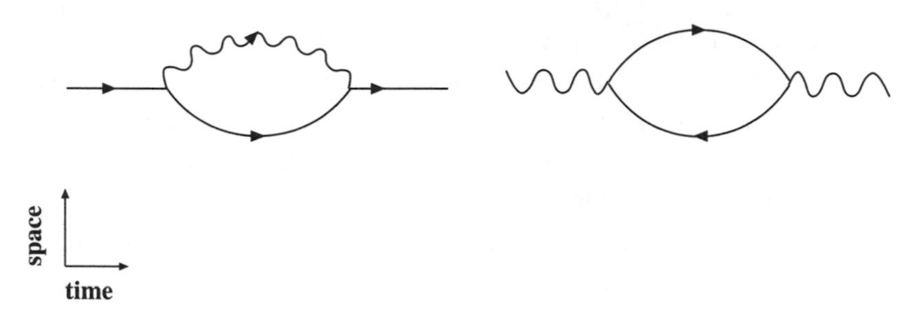

Figure 9.2 Two different loop diagrams. In the left diagram, an electron emits a photon and then reabsorbs it. In the right diagram, a photon decays into an electron-positron pair, which annihilates to form a photon.

scattering equation describes almost 99 percent of what we see. The next few terms account for a correction of almost one percent. Several terms beyond that account for a correction of less than 0.01 percent, and so on. Each term of this series corresponds to a Feynman diagram. The term that accounts for the electron-position virtual pairs, or "loop diagrams" (figures 9.2), as they are called, appears among the "next-to-leading-order" terms, and has the effect of altering high-energy scattering at about the one percent level. This is easily confirmed in the laboratory. The effect of virtual pairs is both measurable and real.

The effect of these virtual pairs not only shows up in electron scattering, but also in the static properties of an electron. The electron has a "self-interaction," which can modify not only such properties as the electron's magnetic moment, but also such common properties as the electron's mass, which we measure in the laboratory. An electron can radiate away a photon and then reabsorb it. Or that photon might decay into a electron-positron pair that recombines into a photon, which in turn is then reabsorbed by the original electron.

But how can we ever really see all this? The answer has to do with scale. If your measurement has a low resolution, the electron and its associated cloud of photons and virtual electron-positron pairs are indistinguishable. A measurement with a coarse resolution is performed on the whole collection. If the measurement has a very fine resolution, however, the field of view might only include part of the system—perhaps only the original electron after it has emitted a photon. In that case it might be a bit shy of its normal energy. But the beauty of QED is that it can take this all in stride and accurately deal with this scale dependence. It can describe the electron as it is measured on any scale.

Returning to the subnucleon world, we might expect QCD to work the same way as QED, and, in principle, it does. Quarks can emit photons and gluons, which can decay into quark-antiquark pairs; the pairs can then be recombined and reabsorbed. What makes QCD so complex are the same factors that made the Roper resonance so curious: the gluon-gluon interaction and the strong coupling between quarks and gluons. It may be a gluon-gluon interaction that makes our picture of the inside of a nucleon a messy web or mesh of gluons, but it is the strength of the coupling that sets the scale of the whole problem.

In QED the inclusion of loop diagrams leads to a correction of only about one percent, but in QCD these loops can give rise to corrections that are nearly as large as the leading-order term of the calculation! This means that a bare quark, collected together with all of its gluons, photons, and virtual quark-antiquark pairs, will look quite different than the bare quark by itself. This collection, in fact, with its mesh of gluons, is a stable, self-sustaining object with its own unique properties. It is this combined system that we call the constituent quark.

So whether we see a bare quark or a constituent quark depends on the magnification of our microscope. If we can resolve objects at a third of a fermi or so, we see constituent quarks and a mesh or collection of gluons—which is sometimes called a flux-tube. If we can resolve objects at a tenth of a fermi, we start to see bare quarks and gluons (see figure 9.3).

So, as a consequence of this complex cloud of virtual quarks and gluons a proton could at a given moment have more than two up quarks. Somehow this is such a reasonable extension of what we readily see in QED that the extra up quark (as long as

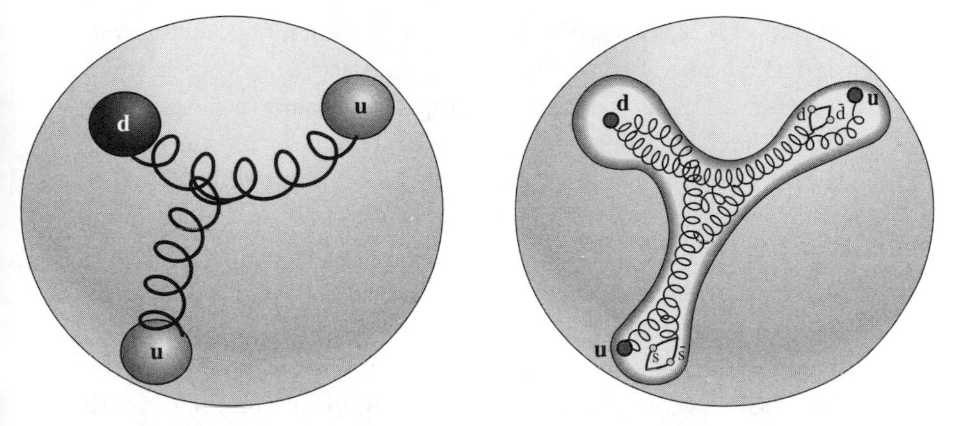

Figure 9.3 If we view a nucleon with a resolution of about a third of a fermi, we see the constituent quarks and flux-tubes. If we view it with a resolution of a tenth of a fermi, we see that it is built up out "bare quarks," gluons, and a number of virtual "bare quark-antiquark" pairs.

it is paired with an antiup quark) will raise no eyebrows and cause no astonishment. However, mention the idea that there are strange quarks within a proton and you can raise quite a stir. When we measure the strangeness of a neutron or a proton we always get zero. This is perfectly consistent with our constituent quark model of the nucleon. But when we dig a little deeper, when we turn up the resolution of our microscope, we see quark-antiquark pairs, and on occasion we see strange-antistrange pairs. Of course we must preserve all those large-scale traits, but since the strange quark is always paired with an antistrange antiquark the macroscopic traits are preserved.

Once there is a wisp of a possibility, the merest suggestion of "strangeness" being buried deep within the most normal of matter, the experimentalist is ready to focus his accelerators and attention on the question. But first let us discuss the likelihood of a strange quark being located in a nucleon. It is true that a gluon passing from a real-and-stable up quark to a real-and-stable down quark has the possibility of disintegrating into a strange-antistrange pair. But the lifetime of that pair is fleeting at best, and that gluon really is much more likely to do something else.

Most likely the gluon will continue on its journey to its targeted real quark and be absorbed without digression or deviation from its itinerary. This is still the "leading order diagram," the most probable fate. The next most likely destiny is for the gluon to decay into an up-antiup or down-antidown pair, but this takes energy. This pair has mass, and we must "borrow" mass from someplace, such as the Heisenberg uncertainty principle. Eventually, the loan must be repaid and the up-antiup pair will collide, annihilate, and produce a gluon again. The loan is more likely to be issued when the amount of energy is small, or if the term of the loan is short. But on occasion the First Bank of the Heisenberg Uncertainty Principle will authorize a more substantial loan, enough to purchase a strange-antistrange pair of quarks. Yet how can we, well outside the nucleon, know that the loan has been issued? It may be interesting to speculate on strangeness in normal matter, but it would appear to be academic at best, and at worst unscientific if we couldn't conceive of an experimental test. However, there have been two types of experiments proposed. One is destructive, the second more subtle, more passive.

The first technique tries to break up the nucleon at the moment it contains a strange-antistrange pair. To do this we would bombard the nucleon with a powerful photon or electron. Most of the time, when it breaks up the split will create a pair of light quark-antiquarks. What we would then see is a nucleon shattered into a pion and a nucleon. But on a rare occasion we could catch the nucleon at the moment it contains the strange pair. When it breaks up one fragment might contain two of the original quarks and a strange quark. This combination is the Λ particle. The remaining original quark and the antistrange antiquark would then combine to form a kaon (K) (see figure 9.4). The reason we need an energetic electron is not only to break up the nucleon, but also because the bank still insists that the energy loan be paid back in full, if not from the energy released in the annihilation, then from the energy of the incoming electron.

An alternative destructive method is to knock out the strange-antistrange pair themselves. The strange pairs form a particle

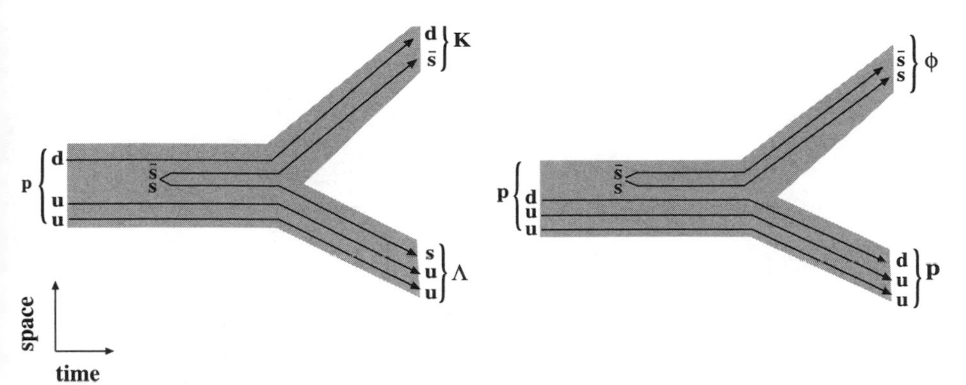

Figure 9.4 If a proton on occasion has a virtual strange-antistrange quark pair, we should be able to break it up at that moment and see the Λ, *K*, or φ, which all contain strangeness.

called a phi (φ). In both of these cases, the lambda-kaon (Λ-*K*) or nucleon-phi (*N*-φ) method, the experiment is simple to describe and relatively simple to perform, but the interpretation of the data is anything but simple. There is always the question, Did the strange pair preexist, or did the virtual photon from the passing electron create them? We can use our old friend spin to help us unravel this enigma, but the signal is not as clean as we would hope.

On the passive, or at least less destructive, side is a unique, but delicate and intricate experiment. This experiment starts out from the question, Do we scatter off a strange quark in a distinctly different manner than off an up or down quark? The electromagnetic and the strong force are "flavor-independent" and do not help us, but that rare force we associate with radioactive decay, the weak force, can offer us a signal.

If we scatter electrons off protons, and they interact via virtual photons, the scattering is symmetric independent of the polarization of the electrons. However, if they interact via virtual Z^0-boson, the scattering is asymmetric. This last statement leaves us with a whole series of questions such as: What is a Z^0-boson? How do we make them? Why is their scattering asymmetric and distinct?

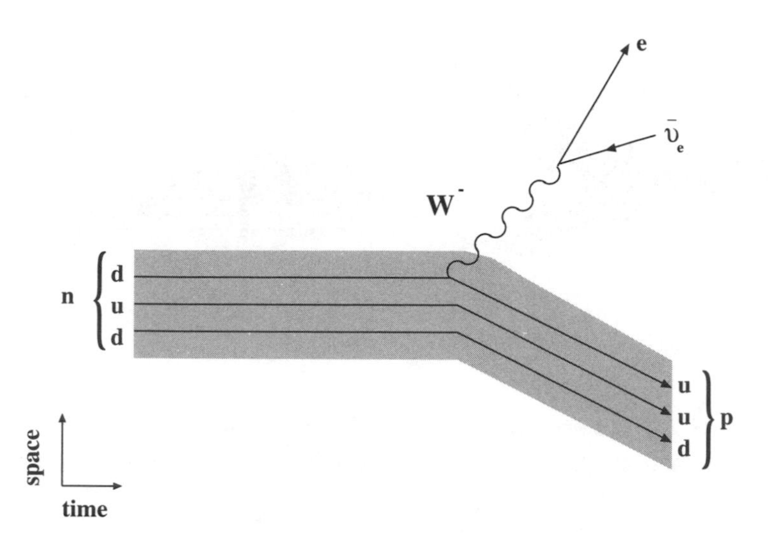

Figure 9.5 In beta decay we do not conserve quark flavor. We started out with two down and one up quarks, and ended with one down and two up quarks. This can only happen in "weak decays."

A *Z*-boson is one of the particles of exchange associated with the "weak interaction." The *Z*-boson is to be compared with the photon for electromagnetic interactions and the gluon for the strong interaction. Curiously enough, the weak interaction has three different exchange particles: the W^+, Z^0, and W^- bosons. The "+," "0," and "–" indicate the charge on these bosons. Actually we have encountered the W-boson already, back in chapter 6 in our discussion of the beta decay and the transmutation of a quark of one flavor into a different flavor. You will remember that in a beta decay a neutron decays into a proton, an electron, and an anti-electron-neutrino. This example of a weak interaction is the prototypical radioactive decay. But what makes this interaction so interesting to our discussion in this chapter is that it has *not* conserved the quark quantum number. Initially we had two down quarks, but by the end we have lost one of them (see figure 9.5)!

This should upset and offend you. Only a few pages ago I said that quark-antiquark pairs could be created only if flavor is conserved. Yet beta decays break that very rule. Let me try to placate you with this brief observation: the Δ particle decays by strong interaction in about 10^{-23} second, whereas the muon decays by weak interaction in 10^{-6} second, and the neutron decays after an expansive and leisurely lifetime of 14 *minutes.* Compared to strong decays, weak decays almost never even happen.

Still, even though the weak interaction is infrequent, it allows peculiar things to happen. First among its peculiarities is its property of "flavor-mixing," that is, allowing a down quark to decay into an up quark. But of greater importance for our purpose is that each flavor interacts with the *W*- and *Z*-bosons in a unique manner. This feature was first manifested in chapter 6, where we saw, for instance, that a charm quark is more likely to decay into a strange quark than into a down or a bottom quark. In most decays in nature the energy that would be released will drive a decay, and so we would expect the charm quark to decay into a down quark and then release the maximum potential energy. But the weak decay's flavor sensitivity changes all of this, and this flavor sensitivity holds for the *Z*-boson as well as the *W*-boson we saw in radioactive decays.

So what is a Z^0-boson? It is neutral, and in many respects it acts like a massive photon. Its mass is 95 GeV/c^2, 100 times the mass of the proton. In fact, its mass is comparable to that of the elements in the middle of the periodic table—somewhere around ruthenium and technetium. Such a large mass means that the Z^0-boson is very short-lived, and so its traveling range is very limited. Nevertheless, it really is there in the heart of electron scattering. When we think of an electron scattering off a nucleon or a quark we usually describe it as an electron coming in and emitting a virtual photon, and it is this virtual photon that strikes the particle. Yet it could have just as easily emitted a Z^0-boson as a photon (see figure 9.6).

Not only is the interaction of the *Z*-boson with each flavor unique, it is also asymmetric with respect to its polarization. In this lies the key to our passive probe. When we scatter *Z*-bosons off a strange quark buried deep within a nucleon, the results are

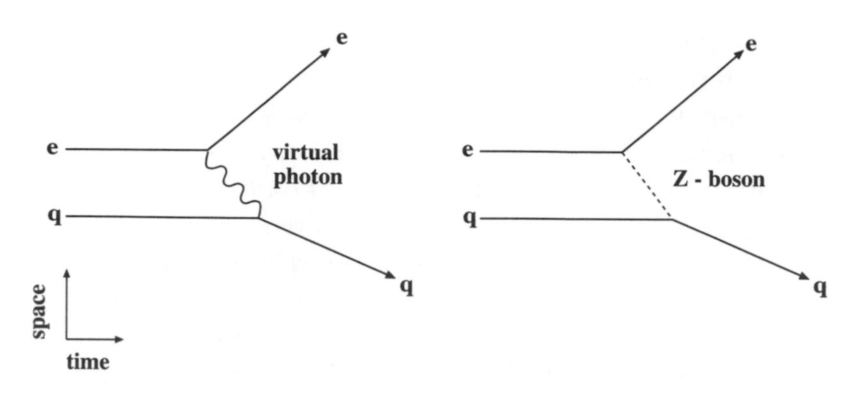

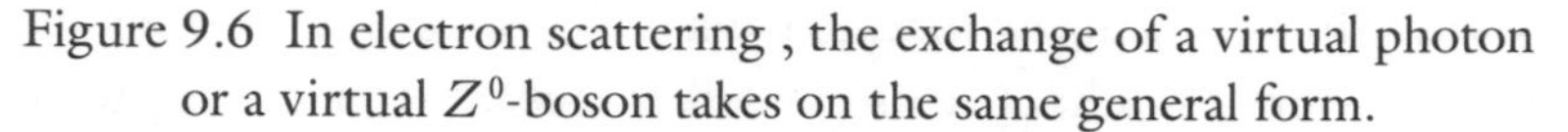
Figure 9.6 In electron scattering , the exchange of a virtual photon or a virtual Z^0-boson takes on the same general form.

slightly asymmetric! The experiment is a difficult one, because the signal is so weak it must be measured in parts per billion (ppb). One has to control the beam's polarization and consistency to an unprecedented degree. Yet this experiment, called SAMPLE, is now under way, just 100 meters from where I am writing these words, at MIT-Bates Lab. The early results are surprising, and like so many other measurements, are leading to a series of follow-up experiments.

If the world was a strange place when we peeled back the surface of a nucleon and peered at the three constituent quarks inside, it is even stranger when we magnify things farther and try to focus our vision on the bare quarks. In some sense we are approaching the gray line that divides nuclear or nucleon physics from high-energy physics, as we start to discuss bare quarks, *W*- and *Z*-bosons, gluons, and strange quarks. But we are still asking the question, What is inside the neutron and the proton? The scope of the question goes beyond just the charge distribution—but that is the way of science: always to dig a little deeper.

- 10 -

A New Age of Exploration within the Hidden World

IF WE HAVE identified all the players in the subnucleon realm and know their "specific distinctions," have we satisfied Carolus Linnaeus's criteria for a scientific study of the field? If we have accelerators that range from a few hundred MeV at NIKHEF, in The Netherlands, to dozens of GeV at DESY, in Germany, is there no longer a direction to grow in?

The answer is No! Linnaeus tells us that naming and sorting is but "the first step of science." In fact, a great many of the experiments described in this book have yet to be performed, and much of the equipment, and even the laboratories, are brand-new or recently refurbished and upgraded. It is the dawn of a new era of scientific exploration. The mapmakers of this microscopic world may have measured the width and length of the ocean, but only a very few islands have been found. Much more detailed and complex questions remain. To continue the geographical analogy, Why do the Hawaiian Islands form a smooth line, from the youngest island, in the southeast, to the oldest island, in the northwest? (The answer: the Earth's crust is moving over a volcanic chimney through which magma rises, building volcanoes and islands.) Why is there a resonance of the nucleon at 1,410 MeV (the Roper), at 1,510, at 1,610, . . . ? I really do not know.

The laboratories of the world are on voyages of exploration. Much as the *H.M.S. Beagle*, *Bounty*, and *Endeavor* plied the waters of the Pacific, the modern nuclear physicist is charting resonances, gathering data, and seeking trends that will help us understand this region. What will we see across these wide expanses? What Galapagoses will challenge our models and our reasoning? So far in this book we have discussed the charge distribution in a nucleon, the nature of the Roper, the search for dibaryons, and the presence of strange quarks inside normal matter.

But that short list is only a sample of the other questions that continue to circulate in the field. For some of them, new data have been taken; for others, no real progress is being made. Is the mass of the neutrino really zero? Recent results seem to tell us it is small, but not exactly zero. Is there something called "color transparency" whereby a quark can travel through a nucleon essentially unaffected? Is the proton truly stable, or can it too decay? Can gluons form a semistable particle called a glueball? The uncharted territory is vast.

One more curiosity has caught the attention of nucleon physicists, but it lies at the high-energy end of the nuclear spectrum. It is a nuclear question that needs a high-energy laboratory to address it. The problem is usually referred to as the "spin crisis."

In the constituent quark model, the spin $\frac{1}{2}$ of a proton or neutron arises from the sum of the spin $\frac{1}{2}$ of the three quarks; the spins of two of the quarks oppose each other and simply cancel, leaving the third quark to account for the spin of the nucleon as a whole. For current quarks the same argument should hold, but at that scale there are other complications: gluons have spin; the collective orbital motions of the quarks and gluons contribute to the nucleon's spin as well. So the question is, How much of the spin of a nucleon really arises from the spin of the current quarks?

Actually making the measurement is very tricky. Physically we can only perform the experiment in a limited range of momentum. We then have to make a number of assumptions to extrapolate to regions where there are no measurements. Still, three major, independent laboratories, SLAC, CERN, and DESY, have all concluded that the three bare quarks together only contribute about 20 percent of the spin of a nucleon! Of course, that finding leads to a flood of new questions: How much of the spin comes from the gluons? How much is due to quarks, how much to antiquarks? How much to up quarks, how much to down quarks? And with every new question, people are imagining new experiments and new methods that might come up with an answer.

New accelerators, new detectors, and even new targets will help explore these questions. At Brookhaven National Laboratory on Long Island is a new accelerator called RHIC—the Relativistic Heavy Ion Collider, which came on-line in 2000. With new detectors and dozens of new experiments it will be looking for evidence of a "quark-gluon plasma." In Sudbury, Ontario, built into an old nickel mine that is the home of the "Big Nickel Monument," is SNO, the Sudbury Neutrino Observatory. SNO will be studying neutrinos from the Sun as well as from the deep cosmos. In Japan is Super-Kamiokande, and in Los Alamos there is LANSCE, the Los Alamos Neutron Science Center.

But just as important as the new apparatuses are the new concepts for probing the hidden world. Even though quarks are as confined today as they were when they were proposed nearly forty years ago, we have developed techniques that probe deep inside nucleons. We "ring" the nucleons and listen for the resonances, or we watch for asymmetries in scattering. We have developed tools such as polarization, and we have learned to infer what we cannot see directly. The confinement of quarks, rather than being the end of physics, is just a tantalizing challenge.

When I gaze into my crystal ball and try to see into the future, my vision is clouded at best. There are some events in the near future that I do see clearly. The charge distribution of a neutron, for instance, should be nailed down in the next few years. But there are a great number of questions that stand off in the more distant future. My vision of them is much more muddled and more obscured. What is the Roper? Are there dibaryons? What is the full role of gluons in a nucleon? I hope I shall live to find out.

Nucleon physics is a young field, and vast territories of it remain unexplored. There is little fear of running out of good questions and interesting problems. Rather, the new generation of experiments, like all good science, seems to raise more questions and curiosities than they answer.

So what does the future hold for the exploration of quarks inside neutrons and protons? I don't know. But I do know that if past experience is any guide, whatever the puzzles and problems, whatever the mysteries, anomalies, and questions encountered along the way, they will satisfy the strongest taste for the exotic, the outlandish, the strange, and the compelling. The people who will map these uncharted regions, both seasoned investigators and new bloods, will bring with them an unusual thirst for delving into a remote corner of the world. Their only guaranteed reward will be to spend a lifetime in a world that is all around us, yet hidden, a land that, like the one Alice found in Wonderland, is "Curiouser and curiouser!"

- GLOSSARY -

α — The fine structure constant, or QED coupling constant, which has the value of approximately 1/137. It helps determine the magnitude of the contribution from various Feynman diagrams in quantum electrodynamics. A first-order diagram, with one vertex, is multiplied by α (~ 0.0073). A second-order diagram, with two vertices, is multiplied by α^2 (~ 0.000053), such that higher-order diagrams have less contribution.

α_s — The strong coupling constant, which has the value of approximately 1/10. This is the coupling constant in QCD. Since it is much larger than α (see above), it means that higher-order diagrams are much more important in QCD than in QED.

4π *detector* — A detector that essentially surrounds a target in all directions. It derives its name from the fact that a sphere one unit in radius has a surface area of 4π units squared.

accelerator — A machine that can push or accelerate particles. Typically in this book we speak of accelerators that can push electrons to nearly the speed of light. Accelerators can also accelerate protons and ions. Since the particles are controlled with magnets, generally accelerators work with charged particles.

ace — The name George Zweig gave to the quarks that he proposed in 1964. The term was derived from the phrase "ace in the hole."

angstrom (Å) — A unit of measurement, 10^{-10} meter, often used in atomic physics. The radius of a hydrogen atom is roughly one angstrom.

antimatter — A particle that has the same mass, but opposite quantum number, as regular matter. For example, the positron is the antimatter partner to the electron. It has the same mass, but opposite charge. The electron charge is $-1e$, the positron charge is $+1e$.

antiquark — The antimatter counterpart to a quark. The up quark, with charge $+\frac{1}{3}e$, is paired with the antiup quark, with a charge of $-\frac{1}{3}e$.

antisymmetric — A system of two or more particles is antisymmetric if by reordering the particles, the quantum number (i.e., spin) of the whole system changes sign. When fermion (spin $\frac{1}{2}$) particles combine, they must form an antisymmetric system (see also the Pauli exclusion principle).

bare quark (or current quark) — A true pointlike quark. This is the most basic quark stripped of its cloud of gluons.

baryon (qqq) — A particle built up out of three quarks. Protons, neutrons, Δ particles, Λ particles, and many more particles are baryons.

Bates Lab — Part of the Massachusetts Institute of Technology, this 1 GeV electron accelerator north of Boston specializes in polarized beams and targets.

beam time — An approved experiment is awarded by a laboratory beam time instead of grant money. Beam time is the number of hours that the laboratory promises to put the accelerator's beam on an experiment's target.

beta decay — The radioactive decay of an isotope that emits an electron. This is the simplest example of a "weak decay."

Bjørken-x variable — A combination of energy loss by the beam and the momentum of a particle detected. Experimentalists can plot their data as functions of the Bjørken-x, and theorists can calculate subnucleon effects in terms of this same variable.

branching ratio — The ratio of two alternative ways a particle can decay. For example, the Λ particle decays into a $p\,\pi^-$ pair 65 percent of the time and a $n\,\pi^0$ pair 34 percent of the time.

Brookhaven National Laboratory — Located on Long Island, east of New York City, Brookhaven is the home of the new accelerator RHIC—the Relativistic Heavy Ion Collider.

calorimeter — A detector that measures the total energy (or calories) of a particle. Typically, it is made of lead glass, heavy crystals, or lead and scintillating plastic. The heavy elements stop the particles; the glass or plastic allows the produced light to be collected in a light detector—usually a photo multiplier tube.

Čerenkov light — Named after the Russian physicist who first explained it, Čerenkov light is caused by a particle moving near the speed of light which enters a medium where the speed of light is much lower. The extra energy and velocity of that particle must be released since it cannot exceed the speed of light in that medium.

color — Quarks carry a "color charge" as well as an electric charge. There is only one electric charge (and it is negative), but there are three color charges—red, green, blue—and their negatives—antired, antigreen, and antiblue. Quarks with the same color repel, and with different color attract. All macroscopic particles are "white," which means that either they are three quarks (baryons) with all three colors, or they are a quark-antiquark pair (mesons) with a color-anticolor charge.

confinement — Quarks are never seen outside a macroscopic particle such as a proton or a pion. At first this was seen as just an experimental fact. Now there is evidence that this may be a theoretical necessity for a "color"-based system.

constituent quark — The effective quark. True quarks (current or bare quarks) are pointlike. But often what we see is related to how these quarks are "dressed" with a cloud of gluons. Dressed quarks, or constituent quarks, form the base of a simple description, and a successful model of what is going on just below the surface of a particle.

constituent quark model (CQM) — A simple but very successful model that is used to describe the types, masses, decay, and structure of particles made out of quarks (hadrons).

cosmic ray — A continuous flux of particles that showers down on us from the upper atmosphere. These are caused by very-high-energy interstellar particles hitting the Earth's upper atmosphere and creating a shower. Most of the particles that reach us are muons, but any other particle can also be found. The first "strange" particles were seen in cosmic ray data. About 100 cosmic rays pass through a horizontal square meter every second.

cross section — The probability of a particle scattering is proportional to its "cross section." Cross section can be thought of as an "effective area," of the target particle.

current quark (bare quark) — A true pointlike quark. This is the most basic quark, stripped of its cloud of gluons.

data silo — A device for storing massive amounts of data. Typically, a data silo is cylindrical in shape, 2 meters tall and 2 meters in diameter. It has hundreds of cubbyholes for tapes facing inside. In the center is a robot arm that can retrieve a tape and put it into a tape reader, when it is needed. Dozens to hundreds of terabytes of data can be stored and retrieved this way.

decay — When a particle decays it "falls apart" into two or more particles. The mass of the daughter particles must be no more than the mass of the mother particle, and is generally less, since the daughter particles can use some of the energy in their momentum.

delta (Δ) particle — There are four delta particles, Δ^-, Δ^0, Δ^+, and Δ^{++}, with a mass of 1,240 MeV/c^2. They are made of three up or down quarks (*ddd*, *ddu*, *duu*, *uuu*), all with their spins in the same direction. This gives them a spin of 3/2 and a 30 percent increase in mass compared to the nucleon.

DESY — The Deutsches Elektronen-Synchrotron, a very large high-energy electron synchrotron facility in Hamburg, Germany.

deuteron — A particle made up of a neutron and a proton. It can exist on its own, or as the core of deuterium, an isotope of hydrogen.

dibaryon — The dibaryon is a theoretically predicted particle that has not been experimentally observed. It is a particle in which six quarks are all bound together into one larger particle.

DoE — The Department of Energy, the primary granting agency for nuclear and high-energy physics in the United States.

electric form factor (G_E) — The electric form factor is a measurement of the charge distribution in a particle. There is a form factor for the proton (G_E^p) and the neutron (G_E^n). The neutron's form factor is of particular interest since it will tell us about the nonsymmetric nature of the up and down quarks in the neutron.

electron — The simplest of leptons. The electron is a pointlike, spin 1/2 fermion. It does not contain quarks. Because it is so simple, is it the perfect tool for probing other particles.

eV, KeV, MeV, GeV, TeV — An eV is an electron Volt. It is a unit of energy equal to the amount of energy an electron will gain when it travels across a one Volt potential. eV are the units used in chemistry. KeV (Kilo electron Volts = 1,000 eV) are the units for X rays. MeV (Mega electron Volts = 1,000,000 eV) are used in nuclear physics. GeV (Giga electron Volts = 1,00,000,000 eV) are used in nuclear and nucleon physics. TeV (Tera electron Volts = 1,00,000,000,000 eV) are used in high-energy physics.

fermi — A unit of measurement, 10^{-15} meter. Nuclear dimensions are typically measured in fermi. Protons and neutrons are a little less than a fermi across.

Feynman diagram — A time-ordered pictorial description of the interaction of particles. Technically, a Feynman diagram can be directly translated into an equation, but sketches like those in this book are also commonly called Feynman diagrams.

flavor — Quarks with different masses are said to have different flavors. These are the different types or species of quarks. There are six flavors: up, down, charm, strange, top, and bottom.

flux-tube — In the constituent quark model, sometimes we treat the web of self-interacting gluons as a "tube" structure. This flux-tube may have dynamic properties of its own, such as vibrational modes.

gluonic excitation — The gluon may possibly have a collective vibrational mode. This has been proposed as an explanation for the Roper resonance.

gluons — Gluons are the particles that convey the strong force between quarks and allow them to interact and bind together.

hadron — Any particle that is made up of quarks and participates in strong (nuclear) interaction. These include all baryons and mesons.

Heisenberg uncertainty relationship — Due to the wave nature of quantum mechanics, there are certain pairs of observables that cannot be measured simultaneously. This leads to a limit on the product of the uncertainties of certain pairs of "observables." These pairs include momentum and position as well as time and energy.

intermediate energy physics — A term sometimes applied to the field of physics between nuclear and high energy.

isospin — Since a neutron and a proton have so many similar characteristics (such as mass and nuclear force), we sometimes treat them as one particle, the nucleon, in two different states. "Isospin" is the label for the differences between these two states.

isotope — Atoms of the same element can have different numbers of neutrons. For example, deuterium is hydrogen with an extra neutron. Since it has one proton and one electron it has the chemistry of common hydrogen, but it has a different mass.

Jefferson Lab — Thomas Jefferson National Accelerator Facility. Located in Newport News, Virginia, this is the newest electron accelerator. It is able to support three experiments simultaneously and achieve energies of up to about 6 GeV.

lepton — The word "lepton" is based on the Greek for "light particle." Leptons are pointlike spin $\frac{1}{2}$ fermions, which do not participate in strong interactions. They include electrons, muons, tau, and the three neutrinos (electron-neutrino, muon-neutrino, and tau-neutrino).

meson — A quark-antiquark bound system. Mesons are the particles that can travel between nucleons to carry the nuclear force. Meson can also have an existence beyond the nucleus. The most common meson is the pion.

muon (μ) — Muons have all the properties of an electron, except that they are about 200 times more massive. They are leptons.

neutrino — Neutrinos are the partners to the electron, muon, and tau particles. Neutrinos are leptons, but since they have no charge and very little, if any, mass, they are hard to detect.

NIKHEF — Nationall Instituut voor Kernfysica en Hoge Energie Fysica in Amsterdam. At one time the home of the AmPS, Amsterdam Pulse Stretcher accelerator.

NSAC — Nuclear Science Advisory Committee, a committee that advises the DoE and the NSF on the most interesting physics to fund.

NSF — The National Science Foundation, the second major source of funding for nuclear/high-energy physics research in the United States after the Department of Energy.

nucleon — Since the proton and the neutron are so similar in many respects, they are sometimes treated as a single type of particle, the nucleon, in two different isospin states.

PAC (Program Advisory Committee) — A committee that advises the head of a laboratory as to the merits of proposed experiments.

partons — A general name for particles that may be used to build a neutron or a proton. The parton model was proposed by Richard Feynman. The particular parton model that best fits the SLAC data is the quark model.

Pauli exclusion principle — The principle that two fermions (spin $\frac{1}{2}$ particles) cannot all have identical quantum numbers. It has been said that the Pauli exclusion principle is what keeps everything from happening in the same place at the same time.

photo multiplier tube (PMT) — A very sensitive light detector. These are often attached to scintillators, Čerenkov detectors, or lead glass calorimeters to convert the light signals into an electrical signal.

photon — A "particle" of light. The photon in the visible range of light will have a few eV of energy. An X ray photon will have

KeV of energy. We can also get photons with MeV or GeV of energy. A photon with about 200 MeV of energy has a wavelength of about the diameter of a proton or neutron. Higher-energy photons are needed to probe the quark structure within these particles.

pion — The most common particle, which travels between nucleons to carry the nuclear force. Pions are also observed outside the nucleus.

polarized — A beam or a target is polarized if the spin axis of the particles within it tend to be aligned in some direction.

positron — The antielectron. It has all the properties of an electron except it has a positive charge. A very energetic photon can decay into an electron-positron pair. Likewise, an electron-positron pair can annihilate and produce an energetic photon.

pseudoscalar meson — A meson (a quark-antiquark system) with spin 0.

quantum chromodynamics (QCD) — The most widely accepted theory of strong interactions. The theory is built on the principle that strong interactions, which hold nucleons and the nucleus together, are a result of the color charges of quarks.

quantum electrodynamics (QED) — Perhaps one of the most precise theories in physics. QED is the theory that describes in detail the interaction of electrons, positrons, and light. It also serves as the model for other quantum field theories.

quark — The most basic fermion that can participate in the strong interaction. Protons, neutrons, and all other hadrons are made up of quarks. Quarks were proposed by Murray Gell-Mann and George Zweig in 1964. We have never seen an isolated quark.

radial excitation — We suspect that a nucleon could be excited by one of its quarks being "kicked" into a higher orbit. This may explain what causes the Roper resonance.

radioactive — An element that spontaneously decays. The decay is caused by the weak force. The most common forms of decay include beta decay (the emission of an electron), alpha decay (an alpha particle is the nucleus of helium-4, two protons, and two neutrons), and gamma decay (the emission of a powerful photon).

resonance — A resonance is seen as a peak in the cross section. This generally indicates that there is enough energy available for a new process, such as the creation of a Δ particle, to take place.

Roper resonance — A resonance seen in the spectrum of electrons scattered from a nucleon. It has a mass of about 1,440 MeV. It may be explained as radial or gluonic excitation, and is a great curiosity.

scattering chamber — The part of an experiment where the beam hits the target. Generally, we are trying to detect what is scattered through this chamber.

scintillator — An organic material, sometimes liquid, sometimes plastic, which produces light when a particle passes through it.

shower counter — A type of calorimeter that measures the energy of a particle by causing it to "shower," usually by colliding with lead and producing a number of lower-energy particles that can be detected with scintillators, or with Čerenkov light.

spectator — A quark or particle is said to be a spectator if it is present in a decay or scattering, but it does not directly take part in the interaction.

spin — Particles have an internal characteristic that we call spin. We often draw these particles like balls that are spinning, even though many of the particles (such as electrons and current or bare quarks) are pointlike and cannot spin in the classical sense. But their spins interact with the same mathematics as balls with classical or macroscopic spin, thus the name.

strangeness — A particle that contains a strange quark is said to have strangeness. Historically, in the 1950s, there were particles that observed in cosmic rays that were described as odd or "strange."

su(3) (special unitary 3) — A set of three objects that obey certain transformation laws. The three light quarks (u, d, s) form an approximate su(3) group. The three colors (red, green, and blue) form an exact su(3) group.

vector meson — A meson (a quark-antiquark system) with total spin 1.

virtual particle — A particle whose existence seems to defy the conservation of energy, but is allowed to exist for a few fleeting moments by the Heisenberg uncertainty relationship.

W^+/W^--boson — A charged particle associated with electro-weak decay. It is massive (80 GeV/c^2) and short-lived. It is associated with radioactive decays.

weak force — The weak force, and associated weak interaction and weak decay, is a few orders of magnitude weaker than the strong or nuclear force. The weak force is what allows radioactive decays. Also, a quark will change its flavor in a weak decay.

wire chamber — A device designed to measure the trajectory of a particle. It is usually a box with thin wires surrounded by a special gas. A charged particle passing through the gas will leave a trail of ions. The ions drift to the wires, where an electrical pulse is induced. By looking at the pattern of pulses from many wires, one can reconstruct the trajectory of the particle that passed through the chamber.

Z^0-boson — A neutral particle associated with electro-weak decay. Like the W^+/W^--boson, it is massive and short-lived.

- INDEX -